Concrete in
Hot Climates

Related RILEM publications

Admixtures for Concrete - Improvement of Properties
Edited by E Vaquez

Chemical admixtures are used to modify the properties and behaviour of fresh and hardened concrete. The can enable concrete to be placed more economically, by reducing the plant and labour needed or by allowing modifications to the mix. They are also used to achieve special properties, such as high strength or freeze-thaw durability. Research and development is underway in many parts of the world, looking at fundamental effects of admixtures on cement hydration and rheology, the technological benefits of using admixtures in concrete, and the long-term behaviour of concrete containing admixtures. This book draws together reports on research studies into the use of chemical admixtures to improve the properties of concrete. It includes papers from leading research institutes and materials specialists from around the world on six main topics: workability, setting, strength, durability, other properties and technology.

 The papers were presented at an International RILEM Symposium held in Barcelona, Spain, in May 1990.

RILEM Proceedings 5, Published 1990, 608 pages, ISBN 0 412 37410 2

Properties of Fresh Concrete
Edited by J H Wierig

The production of concrete has in recent years shifted from the site to central production plants. This has improved efficiency but has also given rise to new problems, such as the influence of temperature and time on consistency of fresh concrete. The separation of mixing and transportation from on-site placing and curing requires better means of measuring and defining the state of the concrete at the time of delivery. In addition, new types of cements and admixtures are being widely used. This book forms the Proceedings of the RILEM Colloquium held in Hanover, Germany in October 1990 to review the state-of-the-art of the properties of fresh concrete. Papers from 18 countries in Europe, North America and the Far East are included.

RILEM Proceedings 10, Published 1990, 400 pages, ISBN 0 412 37430 7

A full list of RILEM publications available from E & FN Spon is given at the back of this book.

Concrete in Hot Climates

Proceedings of the Third International Conference held by RILEM
(The International Union of Testing and Research Laboratories
for Materials and Testing) and organized by RILEM Technical
Committee 94-CHC on Concrete in Hot Climates
and The Concrete Society.

Torquay, England
September 21 - 25, 1992

EDITED BY

M. J. Walker

The Concrete Society, Wexham, Slough, UK

E & FN SPON
An Imprint of Chapman & Hall

London · Glasgow · New York · Tokyo · Melbourne · Madras

**Published by E & FN Spon, an imprint of Chapman & Hall,
2-6 Boundary Row, London SE1 8HN**

Chapman & Hall, 2-6 Boundary Row, London SE1 8HN, UK

Blackie Academic & Professional, Wester Cleddens Road,
Bishopbriggs, Glasgow G64 2NZ, UK

Van Nostrand Reinhold, 115 5th Avenue, New York, NY10003, USA

Chapman & Hall Japan, Thomson Publishing Japan, Hirawacho
Nemoto Building, 6F, 1-7-11 Hirakawa-cho, Chiyoda-ku, Tokyo 102,
Japan

Chapman & Hall Australia, Thomas Nelson Australia, 102 Dodds
Street, South Melbourne, Victoria 3205, Australia

Chapman & Hall India, R. Seshadri, 32 Second Main Road, CIT East,
Madras 600 035, India

First edition 1992

© 1992 RILEM

Printed in Great Britain

ISBN 0 419 18090 7

A catalogue record for this book is available from the British Library.

Library of Congress Cataloging-in-Publication data available.

Publisher's Note
This book has been produced from camera ready copy provided by the
individual contributors in order to make the book available for the Conference.

Contents

Preface

Welcome to the Third International RILEM Conference on Concrete in Hot Climates.

RILEM Technical Committee 94-CHC on Concrete in Hot Climates is staging this conference as part of its project to update and revise the early RILEM work in this field.

The Second RILEM Conference on this topic was held more than 17 years ago and much has happened since. We have discovered that concrete is not quite as durable as desired, particularly in hot climates, and much has been done to improve the situation. Many international, regional and national conferences have been organized, and several excellent handbooks have been published in the intervening period. Some years ago RILEM TC 94-CHC was formed to collect the information and experience on hot weather concreting worldwide, and to publish authoritative guidelines. Before this work is finalized, the Committee would like to discuss its current thoughts with designers, contractors and members of the research community operating in various parts of the world. The Committee felt that such discussion could best be organised in the form of a truly international symposium and it is pleased to welcome you to what should be a well worthwhile and useful examination of the use of concrete in hot climates.

Torben C Hansen
President of RILEM
Chairman of RILEM TC 94-CHC

RILEM Committee Report

The Committee's original intention was to publish their new Report before the conference but the Committee has decided to hold back the completion and publication of its report until after the Conference so that consideration can be given to material presented at Torquay. For each of the Sessions, a keynote presentation will be given by a member of the RILEM Committee to introduce the session topic and to outline the material that will be developed in the RILEM Report. The keynote material is not included in this book but will form part of the material which will be published in the RILEM report in due course.

RILEM Technical Committee 94-CHC
Concrete in Hot Climates

LIST OF MEMBERS

Professor T.C. Hansen (Chairman), Building Materials Laboratory, Technical University of Denmark, Lyngby, Denmark

Dr C. de Fontenay (Secretary), COWIconsult AS, Lyngby, Denmark

Dr H.Z. Al Abidien, Ministry of Public Works and Housing, Riyadh, Saudi Arabia

Dr Z. Berhane, Department of Engineering, Addis Ababa University, Ethiopia

Professor J.M.J.M. Bijen, INTRON, Sittard, The Netherlands

Dr R.D. Browne, Taywood Engineering Limited, Southall, UK

G.R.U. Burg, Master Builders, Cleveland, Ohio, USA

Dr O.Z. Cebeci, Faculty of Engineering, Marmara University, Istanbul, Turkey

Dr N. Cilason, STFA Quality Limited, Istanbul, Turkey

M.E. Ferreira, Building Materials Department, Laboratorio Nacional Engenharia Civil, Lisbon, Portugal

M. Fickelson, RILEM, Cachan, France

Professor P.G. Fookes, Winchester, UK

Professor C. Jaegermann, Building Research Station, Technion City, Haifa, Israel

Dr M.J. Katwan, National Center for Construction Laboratories, Ministry of Housing & Construction, Baghdad, Iraq

G. Macmillan, Materials Testing & Research, Ministry of Works, Power and Water, Manama, Bahrain

N.P. Mailvaganam (Corresponding Member), Building Materials Section, National Research Council, Ottawa, Ontario, Canada

Professor Y. Matsufuji, Department of Architectural Engineering, Kyushu University, Fukuoka-City, Japan

Dr S. Morinaga, Institute of Technology, Shimizu Corporation Limited, Tokyo, Japan

C.A. Ossa, Instituto Columbiano de Productores de Cemento, Bogota, Columbia

Professor Rasheeduzzafar, Department of Civil Engineering, King Fahd University of Petroleum and Minerals, Dhahran, Saudi Arabia

Dr D. Ravina (Corresponding Member), Building Research Station, Technion City, Haifa, Israel

Dr M.A. Samarai, Civil Engineering Department, University of Baghdad, Iraq

Dr A. Samarin, Boral Research, Wentworthville, NSW, Australia

Professor I. Soroka, Building Research Station, Technion City, Haifa, Israel

Dr C.T. Tam, Department of Civil Engineering, National University of Singapore, Singapore

K.W.J. Treadaway, Department of the Environment, London, UK

M.J. Walker, The Concrete Society, Wexham, Slough, UK

Dr S.D. Zivkovic, Faculty of Civil Engineering, University of Belgrade, Belgrade, Yugoslavia

PART ONE

ENVIRONMENT

1 THE EFFECT OF INCREASED TEMPERATURE ON FRESH AND HARDENED CONCRETE

S. D. ZIVKOVIC
Department of Civil Engineering, University of Belgrade,
Yugoslavia

Abstract
The paper deals with the results of an experimental investigation of concrete made of Portland Cement with 15-20% of blastfurnace slag (PCS). In order to establish the effect of increased temperature during concrete works execution, six different temperatures ranging from 8-60°C have been chosen and kept constant during concrete mixing, placing and curing up to the age of 28 days. High relative humidity 95-98% for curing of the specimens in a climate chamber was also kept constant through the ageing.

The development of compressive strength, ultrasonic pulse velocity (UPV), dynamic E-modulus and dynamic Poisson's ratio were tested from a very early stage of concrete hardening (4,8 or 16 hours), up to the age of 28 days. Water absorption and water permeability were also tested at the age of 28 days. The above stated research program was performed with a large number of concrete specimens and gave a huge number of test results.

Only the most important results and conclusions are presented in this paper. The results are mainly presented in the form of diagrams.
Keywords: Increased Temperature, Slump, Workability, Water Demand, Rapid Hydration, Concrete Mix, Strength, E-modulus, UPV.

1 Introduction

The specific problems involved when placing concrete in hot climates are related to fresh concrete, its production process and to concrete of the early ages. These problems arise from three main factors:

(1) An increased rate of evaporation from fresh concrete, during production process and a few hours after its compaction.
(2) A decreased workability of fresh concrete, affected directly by increased temperature of fresh concrete and of the air.
(3) A more rapid hydration of cement after concrete placing and compaction have been finished.

Undesirable effects of the increased rate of evaporation from fresh concrete before placing include an increase of water demand in concrete mixes and an increase in the slump loss. It is, however, possible to control this increase of water demand by field tests of slump loss,

Concrete in Hot Climates. Edited by M. J. Walker. © RILEM
Published by E & F N Spon, 2 - 6 Boundary Row, London SE1 8HN. ISBN 0 419 18090 7.

3

which correspond to the elapsed time and thus enable to keep equal water content at the mixer and at the point of discharge.

Decreased workability i.e. the decreased slump of fresh concrete as a direct effect of the increased temperature of fresh concrete, also includes an increase of water demand in concrete mixes, or otherwise there is a risk of insufficient compacting, honeycombing, cold joints and the like. The following consequences are possible: an increased capillary porosity, if the quantity of cement is kept constant or an increased quantity of cement, in order to keep a constant water-cement ratio i.e. a constant capillary porosity.

A higher capillary porosity leads not only to a lower strength of concrete, but to a lower durability of concrete, too. Increased quantity of cement, however, leads to the increased plastic and drying shrinkage of concrete and the increased risk of crack formation, which also result in a lower durability of concrete structures.

Both of the above stated undesirable effects result in difficulties in handling and finishing the fresh concrete, creating an increase possibility of cold joints or honeycombing, and a grater temptation to add water at the point of discharge.

The more rapid early hydration of cement paste also leads to a lower quality of the hardened concrete, since a less uniform framework of gel and more porous hydration products can be established.

2 Research program and testing procedure

In order to study the effect of the increased temperature during the execution of the works in hot climates on some properties of fresh and hardened concrete, the following experiments have been established:

(a) Measurements of slump of fresh concrete at the constant quantity of water and cement, but at six different temperatures of concrete mix, in the range between 10°C and 58°C.
(b) Determination of water demand for six different temperatures of concrete mix in the range between 10°C and 58°C with a constant quantity of cement and at constant slump of 75±5 mm.
(c) Measurement of the developments of ultrasonic pulse velocity (UPV), compressive strength, dynamic modulus of elasticity and dynamic Poisson's ratio, for six different temperatures in the range between 8°C and 60°C, but for a constant concrete mix. Temperatures were kept constant from concrete mixing up to 28 days, in a climate chamber, with a constant relative humidity of 95–98%.
(d) Measurement of water absorption and water impermeability of the hardened concrete at the age of 28 days, mixed and cured in a climate chamber at four different temperatures: 23, 30, 50 and 60°C, and at a constant humidity of 95–98%.

The basic concrete mix for all tests consisted of 300 kg/m^3 of PC with 15–20% of blastfurnace slag (PCS), 1950 kg/m^3 of the river 'Morava' fine and coarse aggregates, and 180 kg/m^3 of water, except for item (b), where the quantity of water was varied between 180 and 242 kg/m^3.

The slump of fresh concrete measurement, given in items (a) and (b), were performed immediately after concrete discharge from the mixer.

4

Thus, the effect of temperature itself (not of the elapsed time) on slump and on water demand was determined.

The developments of UPV, strength, dynamic E-modulus and dynamic Poisson's ratio were tested from a very early age of concrete hardening, which was 4,8 or 16 hours, depending on concrete mix temperature, then at 1, 2 and 3 days, and 1,2,3 and 4 weeks after concrete mixing. The development of the first two characteristics was tested on the series of three 15 cm cubes for each age at each temperature, while the development of other two characteristics was measured on the series of three 12x12x36 cm prisms for each age at each temperature.

For water absorption and water impermeability tests the series of three 15 by 15 cm cylinders for each temperature have been used at the age of 28 days. Water impermeability test consisted of exposing the lower base of cylinders to 1 bar water pressure and then to the pressure which was for 1 bar higher after each 8 hours of testing. The pressure in bars at which water leakage took place at the upper base of one of the three tested cylinders, was recorded as 'the class of water impermeability'. Water absorption test was performed using the same cylinders. Four temperatures only were applied for the last two tests.

The results of slump measurement and of water demand test as well as the results of the above stated tests were determined as the average values of at least three independent tests.

3   Test results and discussion

The above presented research program and the performed tests resulted in a huge number of results and corresponding analysis, tables, formulas and diagrams. All the details can be found in reference [9]. The most important results only including the corresponding analysis are herein given in several diagrams and a few correlation equations.

3.1   Fresh concrete properties
Fig.1. shows the change of slump of fresh concrete at a constant quantity of water, and the change of water quantity at constant slump, both with the change of fresh concrete temperature. Using regression analysis by the least square method, mathematical approximation of the obtained results are also calculated and shown on the diagram.

It is easy to see from Fig.1. that for fresh concrete temperature increase from 20°C to 40°C a slump decrease of 5.6 cm is recorded and that for the same temperature increase water demand is increased for 24 kg/m³, or by 12.4%. slump decrease and water content increase for the same temperature rise but for concretes made of OPC are, however, considerably lower [1], [2], [9].

3.2   Hardened concrete properties
The development of concrete compressive strength at six different temperatures is shown in Fig.2-a, in log scale for concrete age. The same results, for a better illustration of the temperature effect, are also shown in Fig.2-b, with the compressive strength against mixing -curing temperature, for each particular age of concrete.

It is evident from Fig.2. that the increase of mixing-curing temperature is of benefit for strength development up to 7 days of concrete

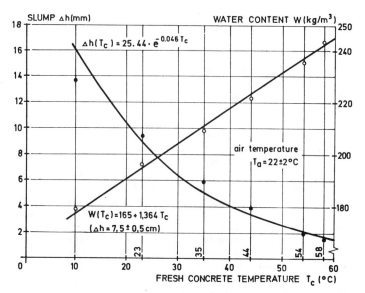

Fig.1. The effect of fresh concrete temperature on slump and water content

age. Moreover, Fig.2-b clearly shows that the temperature increase in the whole range of temperature is of benefit for strength at the earliest age, up to 16 hours only, and that at the ages between 16 hours and 7 days temperature over 40°C does not give any contribution to the strength development. At the ages over 7 days, specimens made of concrete at the lowest temperature of 8 and 20°C have shown the highest compressive strength. For example, 28 days strength of concrete which is mixed and cured at 8°C is more than 40% higher than the strength of the same concrete mixed and cured at 60°C.

The test results of compressive strength development as a function of initial temperature of concrete mixes and of curing temperature enable finding out a mathematical model for description of the effect of temperature on strength development. By the least square method of regression analysis a logarithmic strength-age correlation was previously determined as follows:

$$f_c(t) = a + b \, lnt \qquad (1)$$

where $f_c(t)$ is compressive strength, and t the age of concrete (t≥1day) Correlation constants a and b were calculated for each particular temperature (T). Using the least square method again, a logarithmic correlation for constant a and a linear correlation for constant b against temperature (T) are determined [9].

By taking these relationship a(T) and b(T), and by their substitution in equation (1), the following correlation equation is obtained:

$$f_c(t,T) = -14.15 + 7.06 \, lnT + (9.80 - 0.125 \, T) \, lnT \qquad (2)$$

6

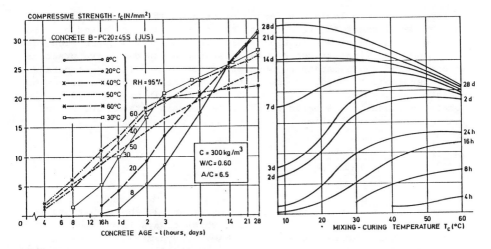

Fig.2. Compressive strength development at six mixing-curing temperatures

where strength $f_c(t,T)$ is obtained in N/mm^2 if age (t) is taken in days and temperature (T=const.) in $^\circ$C. Fig.3 shows the strength-age lines for each particular temperature, together with the points which repre- sent the results obtained by crushing tests. Relative mean deviation (S_R) of calculated against measured values of strength ranging from 3.1% to 9.7%, with the average of 5.3%, that is also shown in Fig.3, corresponds to a very good accuracy of compressive strength estimation.

Fig.4. shows the results of UPV development tests for each particu

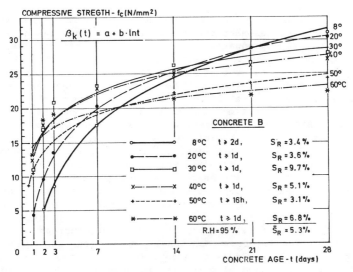

Fig.3. Compressive strength calculated by equation (2) with the points obtained by crushing tests

7

lar temperature, in log scale for concrete age. It is clear from Fig.4. that UPV development at the earliest age is very much affected by temperatures, but the effect on UPV development at later ages is much less pronounced. The final values of UPV, measured at 28 days, are practically unaffected by temperature. Such effect of mixing-curing temperature on UPV development, which is considerably different from the effect on strength development, is explained more detailed in chapter 3.3 but the shortest explanation is based on the relationship between UPV(V) and dynamic modulus of elasticity (E_d):

$$V = \sqrt{\frac{E_d}{\gamma} \frac{1-\mu_d}{(1+\mu_d)(1-2\mu_d)}} \tag{3}$$

where V, γ, E_d and μ_d are UPV, bulk density, dynamic E-modulus and dynamic Poisson's ratio, respectively. Equation (3) shows that UPV is directly affected by dynamic E-modulus, which always increases much faster than the compressive strength. Namely, dynamic E-modulus (as well as static E-modulus) reaches 80-90% of its final value during the first 24 to 48 hours, even at moderate temperature.

Dynamic E-modulus and dynamic Poisson's ratio development, affected by initial and curing temperature, are shown in Fig.5-a and b. Diagrams given in both figures are presented in log-log scale. Values E_d and μ_d are obtained by measurement of resonant frequency (f) of longitudinal vibrations and of ultrasonic pulse velocity (V), on the same concrete prisms. E_d - modulus is calculated by the following equation:

$$E_d = 4 \gamma f^2 l^2 \cdot 10^{-6} \tag{4}$$

where E_d is obtained in N/mm^2, if bulk density (γ) in kg/m^3, length of prism (l) in m, and resonant frequency (f) in 1/s are taken. Poisson's ratio is calculated by equation (3), on the basis of previously determined E-modulus (E_d), pulse velocity (V) and bulk density (γ).

Fig. 5-a. clearly shows that the initial - curing temperature has extremely high effect on the early development of dynamic E-modulus. Concrete mixed and cured at 20 or 30°C reaches respectively 3 times or 10 times higher dynamic E-modulus at 16 hours than concrete mixed and cured at 8°C. Higher temperatures: 40, 50 and 60°C, however, do not give a considerable contribution to the E-modulus, compared to the temperature of 30°C. At later ages, up to 7 days, the differences between E_d-modulus values are decreasing, so that for the ages over 7 days the effect of temperatures becomes opposite: the highest value at 28 days is recorded for concrete mix at 8°C, and the lowest for concrete mix at 60°C. Thus, qualitatively, the influence of temperature is the same as on compressive strength, but there is a considerable quantitative difference: final E_d-modulus values for concrete mixes at 8°C and 60°C differ for 17% from each other, while final strengths for the same temperatures differ more than 40%. The explanation given when UPV development was discussed also covers the development of dynamic E-modulus. It should be emphasized, however, that final values of E-modulus differ slightly more between each other than final values of UPV.

Dynamic Poisson's ratio at early ages, as shown in Fig.5-b, is also

8

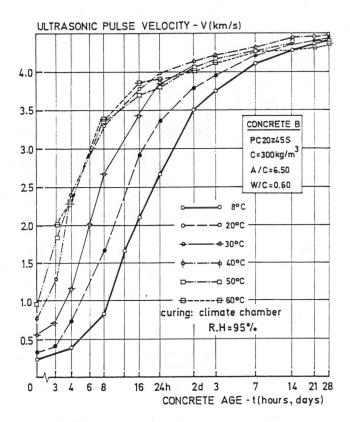

ULTRASONIC PULSE VELOCITY - V(km/s)

CONCRETE B
PC 20z45S
C=300kg/m³
A/C=6.50
W/C=0.60

o————o 8°C
o————o 20°C
o————o 30°C
o————o 40°C
□————□ 50°C
□————□ 60°C
curing: climate chamber
R.H=95%•

CONCRETE AGE - t(hours, days)

Fig.4. UPV development for six mixing-curing temperatures

very much affected by the mixing-curing temperature of concrete. Up to 3 days higher values are obtained when concrete is mixed and cured at lower temperatures. At 16 hours, for instance, values of 0.37 for T=8°C and 0.30 for T=60°C were recorded for Poisson's ratio. At the ages exceeding 3-7 days, up to 28 days, the temperature effect on dynamic Poisson's ratio is changing to opposite direction, as well as has been stated for dynamic E-modulus, and strength of concrete. The lowest value of μ_d=0.255 is recorded for T=8°C, and the highest value of μ_d=0.270 for T=60°C. The difference is lower than for E-modulus.

Finally, the results obtained by testing water absorption and water impermeability are presented in Table 1.

As it is shown in the Table 1, class of water impermeability, defined as the water pressure in bars, at which leakage occured on one of three specimens, is two times lower for concrete mixed and cured at 60°C then for concrete mixed and cured at 20°C. Water absorptions of concrete at 50°C and 60°C are respectively 99% and 90% higher than water absorption of concrete at 20°C.

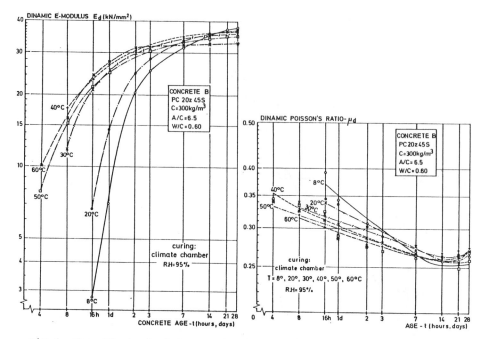

Fig.5. The effect of mixing-curing temperature on dynamic E-modulus (a) and Poisson's ratio (b) development.

Table 1

Initial and curing temperature T (°C)	20	30	50	60
Class of water impermeability (bars)	8	6	4	4
Water absorption (%)	1.76	2.46	3.51	3.33

3.3 The effect of rapid hydration on hardened concrete properties

Not many investigators have experimentally studied the effects of the increased rate of hydration on concrete properties [7], [9]. A few authors, however, have tested the effect of the elevated temperature on the rate of cement paste hydration and thus on hardened cement paste properties [3], [4], [8]. In all sources, however, a higher initial strength and a lower final strength of concrete or cement paste have been obtained. Some of the above stated sources have given theoretical explanation, based on physical and chemical properties of the hydration products. The explanations given by Verbeck and Helmuth [8] then by Neville [6], are the most acceptable, and have been elaborated in details within the reference [9]. Here, a short explanation connected with the results presented in this chapter, will be given.

By the accelerated initial hydration, which occurs at the elevated temperature, no uniform diffusion and the sedimentation of the hydra-

10

tion products in the internal space between the cement grains can be achieved. When cement hydrates at a low temperature, and therefore hydrates slowly, there is an ample time for the hydration products to diffuse and precipitate relatively uniformly throughout the interstitial space, among the cement grains. However, if the hydration is accelerated, as by the increased temperature, the high rate of reaction does not allow time for such diffusion and precipitation. Therefore, a high concentration of hydration products is built up in the zone surrounding the hydrating cement grains. The presence of this highly concentrated, dense and encapsulating layer, which forms an impermeable rim around the cement grain should retard subsequent hydration [8].

It is believed that the dense zone of hydration products around the hydrating grain, created by rapid early hydration, significantly reduces the ultimate degree of hydration of the silicate phases [4], [5].

The hydration products of a weaker physical structure are formed, more porous than usually, so that the larger portion of pores remains unfilled for ever. The hydration level at the later age of concrete is lower, which causes significantly lower ultimate strengths [6].

The results of strength development testing, obtained in the present investigation and shown in Fig.2. and Fig.3, very much coincide with the above stated explanation. Comparing the results obtained in the present investigation with the results obtained by testing concrete made of OPC [5],[9], it is obvious that a negative effect of the increased temperature on the final strength is much more pronounced for concretes made of PCS (PC with blastfurnace slag). It is believed that similar results can be found with Portland Blastfurnace Slag Cement (with a higher percentage of slag).

When the results of testing of UPV (Fig.4.) and of dynamic E-modulus (Fig.5.) are in question, the above stated explanation is confirmed again. The rapid early hydration, affected by the increased temperature, is of benefit for a very high rate of development of E-modulus and of UPV, which is considerably higher than the rate of strength development, even in moderate climate conditions. Because the highest part of full values of E-modulus and UPV is reached at the very early age anyhow, it is clear that at the elevated temperatures, with rapid early hydration, the initial parts of reached E-modulus or UPV values must be even higher. Thus, the effect of rapid hydration cannot very much affect the final values of E-modulus and UPV, because the highest part of these value is already attained at the early age.

Two times lower water impermeability, and two times higher water absorption, registered on concrete at 50°C, and 60°C than on concrete at 20°C (Table 1) have shown the most detrimental effect of high initial temperature on final properties of the hardened concrete. It is clear that the hydration products formed by rapid early hydration are more porous than hydration products formed at a lower temperature. If water permeability and water absorption is higher for concrete hardened at high temperature, it is believed that the durability of this concrete is lower too.

4 Conclusions

On the basis of the results and the discussion presented within the chapter 3 the following conclusions can be established:

(1) During concrete works execution the increased temperature of both freshly mixed concrete and the surrounding air very much influences the fresh concrete workability, the rate of early hydration and the properties of the early aged and hardened concete.

(2) The fresh concrete workability considerably decreases with the increased temperature through rapid evaporation during mixing, transport, discharge and placing and by direct effect of temperature on the concrete mix. In order to enable a satisfactory compaction of the concrete, a higher water-cement ratio, or higher cement content should be applied at elevated temperature,than at moderate climates.

(3) An increased water-cement ratio in concrete mixes leads to a higher capillary porosity of the hardened concete, while the increased cement content, increases plastic and drying shrinkage and therefore the risk of shrinkage cracks formation.

(4) Rapid early hydration, when concrete hydrates at the elevated temperature, additionaly decreases quality of the hardened concrete, since a less uniform framework of gel and more porous hydration products can be established.

(5) Compressive strength at 28 days of concrete mixed and cured at the temperature over $20^{\circ}C$ is always lower than the corresponding strength of concrete mixed and cured at the temperature bellow $20^{\circ}C$.

(6) Water absorption and water permeability of the hardened concrete, mixed and cured at higher temperatures,are also very much increased: at $60^{\circ}C$ both characteristics are two times higher than at $23^{\circ}C$

5 References

[1] Annales (1989) Betonage par Temps Shaud.Connaissances Actuelles et Recommendations. Annales, 474, 79-87.
[2] CIRIA (1984) The CIRIA Guide to Concrete Construction in the Gulf Region, CIRIA Special publication 31, 56-59.
[3] Hansen, P.F. Jessing, J.K. Monsted, R. Trudso, E. (1968) Physical and chemical properties of cement mortar cured at elevated temperatures. International Symposium on Cement, Tokyo.
[4] Idorn, G.M (1968) Hydration of Portland cement paste at high temperature under atmospheric pressure. International Symposium on Cement, Tokyo.
[5] Kasai, Y. Hiraga, T. Yokoyama, K. (1964) Initial strength of concrete (at varied curing temperature), 18th General meeting of JCEA, Tokyo, pp 123-128.
[6] Neville, A.M. (1973) Properties of Concrete. 2. Edition, Pitman Publishing.
[7] Regourd, M. Gauthier, E. (1982) Le durcissement du beton en fonction de la temperature. International Conference on Concrete of Early Ages, Paris, Vol. 1, 145-150.
[8] Verbeck, G.J. Helmuth, R.H. (1968) Structures and physical properties of cement paste. Principal paper. International Symposium on Cement, Tokyo.
[9] Zivkovic, D.S. (1989) Prilog istrazivanju uticaja temperature i drugih relevantnih parametara na neka svojstva svezeg betona i betona male starosti. Thesis of Ph.D. submitted to the University of Belgrade.

2 EVAPORATION OF SURFACE MOISTURE: A PROBLEM IN CONCRETE TECHNOLOGY AND HUMAN PHYSIOLOGY

K. C. HOVER
Department of Structural Engineering, Cornell University,
Ithaca, New York, USA

Abstract
The control of plastic shrinkage cracking, and indeed the entire issue of maintaining a satisfactorily moist concrete surface depends on the rate at which moisture is brought to the surface by whatever means, and the rate at which that moisture is removed from the surface by evaporation. This problem is not unique to concrete construction, however, as the proper functioning of many plants and animals, including the human, depends on maintaining an adequate moisture balance at the skin surface. Human senses can reliably detect changes in the moisture balance, which ability may coincidentally be of value in making qualitative decisions about concrete quality control in the field.

Keywords: Bleeding, Cracking, Curing, Drying, Evaporation, Plastic Shrinkage Cracking, Shrinkage.

1 Introduction

The proposal that there may be useful parallels between evaporative moisture loss in concrete and evaporative moisture loss in plants and animals is supported by the following quotation from the biological literature: "It is axiomatic in contemporary biological thought that living systems require an aqueous internal environment and that water constitutes an indispensable component of cells" [Crowe & Clegg 1973]. This statement could be paraphrased as: "It is axiomatic in current engineering thought that the continued hydration of Portland cement and accompanying development of engineering properties of concrete requires an aqueous internal environment and that water constitutes an indispensable component of the hardened cement paste; or as demonstrated by Powers [1947], hydration of portland cement takes place only in a "water-filled space." Thus it is a mutual dependence on liquid water that unites man and

Concrete in Hot Climates. Edited by M. J. Walker. © RILEM
Published by E & F N Spon, 2 - 6 Boundary Row, London SE1 8HN. ISBN 0 419 18090 7.

concrete within the context of this admittedly unusual paper. Three key points are to be made in this regard:

(a) Maintaining a balance of surface moisture is a key objective for both concrete and humans. Adequate control requires accurate sensing.

(b) Common industrial techniques for measuring the rate at which water appears on the surface of concrete due to bleeding, and the rate at which water disappears due to evaporation are generally inadequate.

(c) By comparison, the human body is highly sensitive to the moisture balance, providing signals in terms of levels of comfort and discomfort in response to the environment and the rate at which moisture is made available to the skin. It appears likely that by learning to recognize a few simple signals, the construction supervisor can use his or her natural senses to aid in making decisions about the care and protection of the concrete.

2 Moisture balance in concrete and man in hot climates

Hot, dry climates can rapidly remove water from the surface of unprotected concrete via evaporation. (Cold, dry climates can be even more desiccating, while hot, moist environments can be relatively benign in regard to drying concrete surfaces.) The concrete itself does not experience the damaging effects of drying as long as surface moisture is provided at a rate that equals or exceeds the rate of evaporation. Such moisture can be supplied by the concrete's own bleed water and/or by fog-spraying in the first few hours after casting, and then by subsequent curing procedures. When the rate of moisture loss exceeds the rate of bleeding and/or fog spraying in freshly cast concrete, crusting, plastic shrinkage, or plastic shrinkage cracking can result. With continued loss of moisture without replenishment by curing, hydration of the cement is inhibited with permanently debilitating effects in the zones affected by desiccation. A given concrete mix can lose only a finite amount of water to the environment before its health is permanently impaired.

Likewise, the human body depends on water for survival, and "health" requires the replenishment of water lost to the environment. Just as in concrete, the loss of surface moisture from the skin at a rate faster than the rate at which water can be transported from the underlying tissues results in drying, crusting, chapping, cracking, and peeling [Blank 1952, Gaul 1952, Lee 1964]. Continued unchecked moisture loss from the human body then results in thirst, dehydration, and permanent injury [Adolph 1947, Kirmiz 1962, Schmidt-Nielsen 1964].

14

Complicating the analogy between concrete technology and human physiology in the hot environment, however, is the fact that while one can preserve concrete quality by preventing evaporation altogether, human survival in hot climates requires the evaporative cooling of perspiration to remove excess metabolic heat and thus maintain a near constant body temperature. Human comfort in a hot climate therefore demands an active rate of evaporation from the skin surface. This is useful, however, as the level of human comfort under hot, dry conditions is an index of the rate of evaporation [Fanger 1970/72]. In contrast, concrete "comfort" in a hot climate demands a limited rate of evaporation from the surface.

In both the case of concrete and the human, therefore, it is critical that water loss be balanced by water replacement, and in each case "injury" can result from failure to maintain the balance. Steps to ensure healthy concrete and healthy people include monitoring the rates at which water is supplied and lost, and replenishing the supply at frequent intervals.

3 Attempts to control moisture balance in concrete

3.1 General

Menzel [1954] initially proposed that risk of plastic shrinkage cracking develops whenever the rate of evaporation at the surface of the concrete exceeds the rate of bleeding. To use this proposition to prevent plastic cracking in the field, one either takes steps to reduce the rate of evaporation to less than the bleeding rate, or one avoids concrete placement altogether when evaporative conditions are too severe. (Lerch [1957] demonstrated that altering the mix to increase bleeding rate is not effective.) Several public and private agencies in the U.S. have specified maximum rates of evaporation which may develop on site beyond which the concreting operations will be terminated.

Quantitative control of plastic shrinkage cracking would therefore include a rational measurement of the rate of bleeding to determine the tolerable rate of evaporation, followed by accurate measurement of the rate of evaporation. This has turned out to be more accurately said than done, as the state-of-the-practice in the U.S. and evidently elsewhere, is to **assume** a value for tolerable evaporation rate and then to estimate evaporation in a manner different from that intended by the developer of the most commonly used method.

Both the rate of bleeding and the rate of evaporation are mix-specific, site-specific, time-dependent, and measurable variables. Bielak and Nicolay [Bielak 1990, Nicolay 1990] have recently demonstrated useful field tests for determining the bleeding rate of concrete, and alternative tests such as ASTM Method C 232 have been available for some time. Likewise, due to its importance in the fields of water resources, agriculture, and biology, multiple techniques are available for determining rate of evaporation [Brutsaert 1984]. Menzel [1954] suggested one of these many available relationships for predicting evaporation rate for concrete and construction applications. Multiple techniques are therefore available to measure rate of bleeding and rate of evaporation for a given mix and set of site conditions; a rational comparison and decision-making process **could** follow.

3.2 Assessing bleeding rate

Bleeding measurements are not normally made for the purpose of assessing the tolerable rate of evaporation, however. In fact, the bleeding rate is not normally associated with the problem of plastic shrinkage among the rank and file in modern practice for the following reason. After Menzel and Lerch published their several papers discussing the relationship among bleeding, evaporation, and plastic shrinkage, the National Ready Mixed Concrete Association in the U.S. published an information bulletin summarizing the previous work and discussing that while bleeding rates may vary between 0.5 to 1.5 kg water/m^2/hr evaporation rates above about 1 kg/m^2/hr "make institution of precautionary measures almost mandatory [NRMCA 1960]." This caveat, with the cautionary criterion at about 1 kg/m^2/hr [0.2 lbs/ft^2/hr] became firmly embedded in U.S. practice, ultimately losing its original association with bleeding rate. Current recommendations [ACI 305 1989, ACI 308 1991, FIP 1986] generally neglect the direct association with bleeding, citing tolerable evaporation rates of between 0.5 and 1.0 kg/m^2/hr as apparently applicable to all concrete mixes.

The primary difficulty here is of course that bleeding rate varies widely as a function of mix characteristics and member geometry [Menzel 1954, NRMCA 1960, Bielak 1990, Nicolay 1990, Schiessl 1990, Suhr 1990]. Reported rates have varied between almost 0 kg water/m^2/hr, up to about 1.5 kg/m^2/hr. (The use of fine cements and especially microsilica depresses the bleeding rate to the near zero level.) As a consequence mixes and members vary considerably in the rate of evaporation that can be tolerated. Most notably those mixes with essentially zero bleeding rate, (a thin bridge deck overlay employing a microsilica concrete, for example) can tolerate essentially zero evaporation; i.e., surface drying begins immediately upon placement. An evaporation-rate criterion permitting up 0.5 to 1.0 kg/m^2/hr is meaningless in such cases, and zero evaporation must be maintained by means of continuous fog spraying. Thus, the initial problem is the failure to

16

associate tolerable level of evaporation with actual rate
of bleeding. This is followed by the companion problem of
inappropriately measuring the rate of evaporation.

3.3 Assessing rate of evaporation
It is conventional practice to calculate the evaporation
rate rather than to measure it, through the use of Menzel's
predictive equation mentioned above. This equation
requires values for concrete temperature, air temperature,
relative humidity, and wind speed as input. The predictive
equation itself is rarely used, however, as the graphical
solution to the mathematical expression, known as "the
evaporation nomograph" has become internationally familiar
[ACI 305 1989, ACI 308 1991, FIP 1986, Murdock et. al.,
1991]. To understand the limitations and common errors
associated with the use of Menzel's equation, however, it
is useful to briefly trace its background.
Menzel's equation originates in "Dalton's Law" [Dalton
1802] proposed in 1802, based on the fundamental physics of
the evaporation process. Dalton postulated the following:

$E = f(V) \ (P_{ws} - P_{wa})$ where,
E = rate of evaporation in units of height of water
per unit time.
P_{ws} = saturation water vapor pressure of liquid at the
temperature of the evaporating surface.
P_{wa} = water vapor pressure of the environment.
$f(V)$ an undefined function of wind velocity, V and the
geometry of the evaporating surface.

According to Brutsaert [1984] "countless" subsequent
researchers developed their own empirically determined
versions of the wind speed function, and Menzel's
relationship is just one more variation on Dalton's theme.
Note that each of these variations, Menzel's included, was
intended as a means of predicting evaporation of water from
a free water surface, directly applicable to predicting
evaporation from bodies of water such as lakes, streams,
and reservoirs, or for predicting the rate of evaporation
of bleed water from a **wet** concrete surface.
As empirically determined relationships, all such
expressions are valid only within the limits of the
experimental conditions. Menzel's equation is no
exception, with the specific rules for obtaining the
requisite input data accompanying his original work as
follows [Menzel 1954]:

17

$W = [0.44(0.253 + 0.096V)](e_o-e_a)^*$, where

    W = lb. of water evaporated per sq.ft. [ft^2] of
surface per hour.
    e_o = pressure of saturated vapor in lb. per sq.in.
[in^2] at temperature of evaporating surface.
    e_a = vapor pressure of the air in lb. per sq.in.
[in^2]. This can be obtained from dry bulb and wet-
bulb temperatures to give the dew-point temperature
of the air. For this purpose dry-and wet-bulb
temperature should be measured at a level about 4 to 6
ft. [1.3 to 2 m] higher than the evaporating surface
on its windward side and shielded from the sun's rays.
    V = average horizontal air or wind speed in miles per
hour. This should be measured at a level about 20 in.
[0.5 m] higher than the evaporating surface.

* SI Equivalent = $E = [0.04(0.253 + 0.06V)](e_o-e_a)$;
V in km/hr; e in mm Hg.

    Menzel published a nomograph for solving this equation
in 1954, a rather complicated version which was revised 6
years later in a publication by the National Ready Mixed
Concrete Association [NRMCA 1960]. This improved version
was easier to use, employed temperature and relative
humidity rather than vapor pressure, and subsequently
became adopted internationally as "The Evaporation
Nomograph." Unfortunately, this popularized version
deleted the guidance for obtaining the necessary input data
and these critical instructions have been missing ever
since.
    Whereas Menzel's empirical constants were based on
average wind velocity measured at 0.5 m above the
evaporative surface, many practitioners enter the nomograph
with peak or gusting wind speeds reported from the nearest
weather station or airport control tower. Since wind
velocity can vary significantly with relatively small
variations in height above the ground [McDonald 1975,
Sachs 1978] entering the nomograph with a wind speed value
obtained at other than 0.5 m above the evaporating surface
can lead to errors on the order of 100% of the computed
value. Failure to meet Menzel's qualifications for
obtaining temperature and humidity values can lead to
errors in calculated evaporation rate as well, although
these effects are not as significant as that of the wind
speed. Shaeles [1988] has documented the error in
inappropriately using Menzel's equation.

4 Human sensibility

4.1 General

When computing evaporation rate by the Menzel formula, it is interesting to note that the result can be grossly in error without such error being obvious to the person performing the calculation. This may be because the process of making the measurements and performing the calculation is rather non-intuitive, and the person involved may have little "feel" for the correctness of a predicted evaporation rate in terms of $kg/m^2/hr$ or $lbs/ft^2/hr$. That the human could misjudge the evaporation rate is somewhat ironic, because the human body is itself highly sensitive to rate of evaporation, and as discussed above the level of human comfort in a hot climate is directly related to evaporation rate. (An in-depth exploration of the physiological issues here is well beyond the author's expertise and the scope of this paper. Nevertheless, some basic relationships of practical utility can be demonstrated, and those with deeper interest are invited to explore the references to this paper.)

The ability to sense evaporation rate arises from the body's acute sensitivity to the loss or gain of heat as influenced by the temperature of the skin, the temperature of the air, and the velocity and relative humidity of the air [Siple 1945, Fanger 1970/72, Precht 1973, Darian-Smith 1984]. These are of course the terms which define rate of evaporation as described by Dalton, and variations on his law of evaporation are used in human physiology to predict rate of evaporation from the surface of the skin [Blank 1952, Gaul 1952, Tregear 1966, Precht 1973, Boutilier 1979].

4.1 Perspiration and Evaporation

The evaporation of a gram of perspiration from the surface of the skin removes approximately 500 cal of heat energy from the human body [Schmidt-Nielsen 1964]. This mechanism is the primary means of regulating internal body temperature in hot environments. (The perspiration reaction is triggered at ambient temperatures above about $30^{\circ}C$ [Blank 1952, Kirmiz 1962].) When a human is working in a hot environment, the rate of production of perspiration can be in the range of 0 to 2 kg/m^2 of skin surface/hr, depending on environmental conditions and the level of activity [Kirmiz 1962, Schmidt-Nielsen 1964, Tregear 1966, Adolph 1947]. At typical (not peak) levels of exertion in air temperatures between 30 and $40^{\circ}C$, construction workers may perspire on the order of 1 $kg/m^2/hr$, while those observing from the shade may only be perspiring at say 1/4 of that amount [based on Adolph's work with soldiers in the desert].

When the rate at which perspiration arrives at the skin surface is greater than the rate of evaporation, the perspiration is "sensible," i.e., it is both felt and seen on the skin and is absorbed into the clothing. When the rate of evaporation matches or exceeds the rate of perspiration, not only is the perspiration not seen or felt, the result is an adequate level of evaporative cooling and the feeling of comfort. Those who extol the virtues of living in the desert by saying, "but its a dry heat," are reflecting on the fact that the rate of evaporation exceeds the rate of perspiration.

This physiological background is related to concrete via the fact that the level of human comfort in a hot climate is directly related to the evaporation rate in comparison to the rate of perspiration. Whereas concrete is "comfortable" when evaporation is equal to or less than the rate of bleeding, the human is comfortable when the evaporation rate is equal to or greater than the rate of perspiration. It is an interesting coincidence that typical human perspiration rates in hot climates are in the same approximate range but somewhat higher than typical bleeding rates for most concretes. Similarly, evaporation rates high enough to make the human comfortable are almost certainly higher than the bleeding rate of most concretes. If the workers are comfortable in a hot environment, it is almost certain that the concrete is at risk of drying on the surface. Alternatively, if the humidity is high in hot weather and the wind speed is low, the workers on site will be uncomfortable, perspiration will be profuse and highly visible on skin and clothes, but the risk of drying of the concrete is much reduced.

4.2 Water loss-thirst, dehydration, and injury.
Continued loss of water from the human body without replenishment leads to injury in the human just as it does in concrete. Concrete and man are similar also in that dangerous amounts of water can be lost prior to signals that water is required.

One of the first signs of water loss in humans is an increase in deep body temperature, at a rate of about 1/2 degree C for every 1% loss in body weight as water [Kirmiz 1962]. While thirst can begin at 1 to 2% loss in body weight, man is observed to not drink enough water voluntarily such that a 2 to 4% deficit can be developed until thirst is compelling [Schmidt-Nielsen 1964]. Thirst is not only a delayed reaction, but it can be satisfied with an intake of only 50% of the water deficit.

Ill effects and functional difficulty are noticed at about 5% loss, or only slightly higher than the level of compelling thirst. Life is endangered at 10 to 12% loss, and the survival limit is a 20% loss in body weight due to water loss [Kirmiz 1962, Schmidt-Nielsen 1964].

To relate these figures to concrete we must first put the quantities of water-loss in comparative units. If the human becomes thirsty at a water loss of say 3% of body weight, this translates to about 2 kg of water loss for a 70 kg person. Over a body surface area of about 1.7 m^2 [Schmidt-Nielsen 1964], thirst may be associated with a cumulative water loss of about 1.2 kg/m^2. This value is in the range of deleterious values of water loss in concrete, such as may result from an hour of evaporation at 1.2 kg/m2/hr, or two hours at half that rate. This suggests that under hot, dry site conditions in which copious perspiration is not visible on skin or clothing but the workers are nevertheless voluntarily drinking water while placing and finishing the concrete, the chances are good that the rate of evaporation is high enough to place the concrete at risk of drying.

4.3 Dry skin and chapping
Finally, a strong analogy exists between crusting and plastic shrinkage cracking of concrete and the drying, chapping, and cracking of human skin. This is because both phenomena are caused by the unreplenished loss of surface moisture, leading to a dry, embrittled surface which cracks on movement [Menzel 1954, Lerch 1955, Blank 1952, Rothman 1954]. Blank's words in the Journal of Investigative Dermatology could explain either phenomena: "Under a wide range of environmental conditions, water is lost from the surface more rapidly than it reaches the surface."
The face and particularly the lips are among the most sensitive areas for this type of drying, manifested in the sensations of chapped lips or wind-burned skin. (These sensations can often be more intense after getting out of the wind.) While quantitative relationships are not available, it appears reasonable to merely associate these sensations with high rates of water loss, and to be concerned about protecting the concrete whenever it is appropriate to protect the skin. When those on site reach into their pockets to take out the familiar tube of chap-stick or lip-balm, it is a good indicator that environmental conditions are hostile to the concrete as well.

5 Summary

Maintaining a balance of surface moisture is a key objective for both concrete and humans. Adequate control requires accurate sensing. The common industrial techniques for measuring the rate at which water appears on the surface of concrete due to bleeding, and the rate at which water disappears due to evaporation are generally inadequate, however. Improvements need to be made in evaluating tolerable and actual evaporation rates, but new scientific advances are not necessarily required. Attention to detail in the use of current methods may be satisfactory for practical purposes.

By comparison, the human body is highly sensitive to the moisture balance, providing signals in terms of levels of comfort and discomfort in response to the environment and the rate at which moisture is made available to the skin. It appears likely that by learning to recognize a few simple signals, the construction supervisor can use his or her natural senses to aid in making decisions about the care and protection of the concrete. The following decision rules are tentatively suggested:

(a) In hot, dry climates, the risk of drying of the concrete is inversely proportional to the level of human comfort. If workers are comfortable, the concrete probably is not. A "dry heat" may be comfortable to the human but is hostile to the concrete.

(b) In hot, dry climates where copious perspiration is not apparent, thirsty workers signifies a thirsty concrete. This rule does not apply in hot, moist environments where thirst is induced by perspiration not removed by evaporation.

(c) When returning to the field office, the warm, stinging sensation described as "wind burn" is the result of rapid evaporation. Concrete cast in weather that induces this sensation is at risk of rapid drying as well. Similarly, when the environment causes those on site to put on lip-balm, the concrete is at risk of rapid drying.

References

Adolph, E.F., *Physiology of Man in the Desert*, Interscience Publishers, New York, 1947, 357 pp.

American Concrete Institute, "Standard Practice for Curing Concrete, (ACI 308-81, Revised 1991)" ACI Committee 308, American Concrete Institute, Detroit, 1991.

American Concrete Institute, "Hot Weather Concreting, (ACI 305R-89)" ACI Committee 305, American Concrete Institute, Detroit, 1989.

Bielak, E., "Testing of Cement, Cement Paste and Concrete, Including Bleeding. Part 1: Laboratory Test Methods," *Properties of Fresh Concrete*, Procedings of the RILEM Colloquium, Hanover, West Germany, October 3-5, 1990, H.J. Wierig, Editor, Chapman and Hall, London, 1990, pp. 154-166.

Blank, I.H., "Factors Which Influence the Water Content of the Stratum Corneium," *Journal of Investigative Dermatology*, Vol. 18., 1952, pp. 433-440.

Boutelier, C., "Survival and Protection of Aircrew in the Event of Accidental Immersion in Cold Water," AGARDograph, No. 211 (Eng.), North Atlantic Treaty Organization Advisory Group for Aerospace Research and Development, February, 1979, 118 pp.

Brutsaert, W., *Evaporation into the Atmosphere, Theory, History, and Applications*, Kulwer Academic Publishers, Dordrecht, 1984, 299 pp.

Crowe, J.H., Clegg, J.S., *Anhydrobiosis*, Dowden, Hutchinson, and Ross, Inc., Stroudsburg, PA, 1973, 477 pp.

Dalton, J., "Experimental Essays on the Constituents of Mixed Gases," Mem., Manchester Lit. and Phil. Soc., Vol. 5, pp 535-602.

Darian-Smith, I., "Thermal Sensibility," *Handbook of Physiology*, Section I: The Nervous System, Vol. III Sensory Processes, Part 2, Brookhart and Mountcastle, Editors, American Physiological Society, Bethesda, Maryland, 1984, pp 879-913.

Fanger, P.O., *Thermal Comfort*, Danish Technical Press, Copenhagen, 1970.

Fanger, P.O., *Thermal Comfort, Analysis and Applications in Environmental Engineering*, McGraw Hill, 1972, 244 pp.

Federation Internationale de la Precontrainte, *Concrete Construction in Hot Weather*, FIP Guide to Good Practice, Thomas Telford, London, 1986, 16 pp.

Gaul, L.E., Underwood, G.B., "Relation sf Dew Point and Barometric Pressure to Chapping of Normal Skin," *Journal of Investigative Dermatology*, Vol. 19, 1952, pp. 9-19.

Kirmiz, J.P., *Adaptation to Desert Environment*, Butterworths, London, 1962, 168 pp.

Lerch, W., "Plastic Shrinkage," Journal of the American Concrete Institute, *Proceedings*, Vol. 53, 1957, pp 797-802.

Lee, D.H.K., "Terrestrial Animals in Dry Heat: Man in the Desert," *Handbook of Physiology--Adaptation to the Environment*, D.B. Dill, E.F. Adolph, C.G. Wilber, Editors, American Physiological Society, Washington, 1964, Section 4, pp. 551-582.

McDonald, A., *Wind Loading on Buildings*, Applied Science Publishers, John Wiley & Sons, New York, 1975, 219 pp.

Menzel, C.A., "Causes and Prevention of Crack Development in Plastic Concrete," *Proceedings* of the Portland Cement Association, Annual Meeting, 1954, pp 130-136.

Murdock, L.J., Brook, K.M., Dewar, J.D., *Concrete Materials and Practice, 6th Edition*, Edward Arnold Publishers, London, 1991, 470 pp.

National Ready Mixed Concrete Association (NRMCA), "Plastic Cracking of Concrete," *Engineering Information*, July, 1960

Nicolay, J., "Testing of Cement, Cement Paste and Concrete, Including Bleeding. Part 2: Field Observations and Their Relation to Laboratory Tests," *Properties of Fresh Concrete*, Procedings of the RILEM Colloquium, Hanover, West Germany, October 3-5, 1990, H.J. Wierig, Editor, Chapman and Hall, London, 1990, pp. 167-186.

Portland Cement Association, "Prevention of Plastic Cracking in Concrete," *Concrete Information*, Structural Bureau, ST 80, 1955.

Powers, T.C., Brownyard, T.L., "Studies of the Physical Properties of Hardened Cement Pastes, Parts 1-9," ACI Journal, *Proceedings* Vol. 43, Oct 1947-April 1948.

Precht, H., Christophersen, J., Hensel, H., Larcher, W., *Temperature and Life*, Springer-Verlag, New York, 1973, 779 pp.

Rothman, S., *Physiology and Biochemistry of the Skin*, University of Chicago Press, 1954, 741 pp.

Sachs, P., *Wind Forces in Engineering, 2nd Edition*, Pergamon Press, Oxford, 1978, 400 pp.

Shaeles, C.A., Hover, K.C., "The Influence of Construction Operations on the Plastic Shrinkage Cracking of Thin Slabs," ACI Materials Journal, Nov.-Dec. 1988, pp 495-504.

Schiessl, P., Schmidt, R., "Bleeding of Concrete," *Properties of Fresh Concrete*, Procedings of the RILEM Colloquium, Hanover, West Germany, October 3-5, 1990, H.J. Wierig, Editor, Chapman and Hall, London, 1990, pp. 24-32.

Schmidt-Nielsen, K., *Desert Animals, Physiological Problems of Heat and Water*, Oxford Press, London, 1964, 277 pp.

Siple, P.A., Passel, L.F., "Measurements of Dry Atmospheric Cooling in Sub-Freezing Temperatures, *Proceedings*, American Philosophical Society, Vol. 89, 1945, pp. 177-199.

Suhr, S., Schoner, W., "Bleeding of Cement Pastes," *Properties of Fresh Concrete*, Procedings of the RILEM Colloquium, Hanover, West Germany, October 3-5, 1990, H.J. Wierig, Editor, Chapman and Hall, London, 1990, pp. 33-40.

Tregear, R.T., *Physical Functions of Skin*, Academic Press, London, 1966, 185 pp.

3 ESTIMATION OF EVAPORATION FROM CONCRETE SURFACES

O. Z. CEBECI and A. M. SAATCI
Faculty of Engineering, Marmara University, Goztepe,
Istanbul, Turkey

Abstract
Evaporation has worried hydrologists long before hot
weather concrete technologists observed severe cracking
on exposed concrete surfaces. On the other hand, process
engineers involved in drying have elaborated on the
fundamentals of evaporation in order to maximize the
efficiency of drying processes. Concrete technologists
have used the results of these studies and have adopted a
nomograph for predicting the rate of evaporation from
concrete surfaces. High rates are warnings of possible
plastic shrinkage cracking. However, recent elaborations
have challenged indiscriminate use of this approach. In
the present paper the methodologies of hydrologists and
process engineers are briefly explained and their data
and results are compared. These methods do not yield a
unique estimate. Finally, a review of concrete
technologists approaches and recent reservations
regarding the estimation of the rate of evaporation from
concrete in hot climatic conditions is given. It is
concluded that collection of more accurate and relevant
data on the rate of evaporation from concrete in hot
climatic conditions and its effect on plastic shrinkage
cracking is necessary so that hot weather concreting
procedures and precautions can be refined.
Keywords: Concrete, Dalton's Law, Drying, Evaporation,
Hot Climate, Menzel's Formula, Plastic Shrinkage
Cracking.

1 Introduction

The word "evaporation" is derived from "vapor" and a
dictionary definition of evaporation is "the act or
process of converting or being converted from a solid or
liquid state into a vapor or gas". This conversion
requires energy known as latent heat of vaporization. One
gram of water requires 540 calories to change into vapor.

Concrete in Hot Climates. Edited by M. J. Walker. © RILEM
Published by E & F N Spon, 2 - 6 Boundary Row, London SE1 8HN. ISBN 0 419 18090 7.

Evaporation from exposed free water surfaces naturally occurs since there is an input of energy either directly from the sun (radiation) or indirectly from the atmosphere itself (conduction from the overlying air). Evaporation can also take place at the expense of energy stored or generated below the evaporating surface.

Evaporation has worried hydrologists long before hot weather concrete technologists observed severe cracking on exposed concrete surfaces. On the other hand, process engineers involved in drying have elaborated on the fundamentals of evaporation in order to maximize the efficiency of drying processes. Concrete technologists have used the results of these studies and have adopted a nomograph for predicting the rate of evaporation from concrete surfaces. High rates are warnings of possible plastic shrinkage cracking. However, recent elaborations have challenged indiscriminate use of this approach. In the present paper the methodologies of hydrologists and process engineers are briefly explained and their data and results are compared. Concrete technologists approach and recent reservations are reviewed and recommendations are given.

2   Rate of Evaporation

The rate of evaporation is controlled by the rate at which the vapor produced can diffuse away from the evaporating surface which is strongly affected by the velocity and flow characteristics and saturation (relative humidity) deficit of the air and surface characteristics of the evaporating body.

The basic components of the evaporation process from a free water surface were first expressed in quantitative terms by Dalton in 1802, who suggested that if other factors remain constant, evaporation is proportional to the windspeed and saturation deficit, i.e., the differences in the vapor pressures at the water surface and in the atmosphere. In practice other factors do not remain constant and a number of meteorological and physical factors affect the rate of free water evaporation (Perry et al., 1984, Ward and Robinson, 1990).

Dalton's law, although never expressed by him in mathematical terms, has formed the starting point of much of the subsequent work on evaporation. Numerous different equations and nomographs for estimating evaporation from free water surfaces have been developed. Although many are based on Dalton's law they do not yield a unique estimate due to the variation in the meteorological and physical factors prevailing the geographical and experimental conditions of the different studies. Meteorological factors include solar radiation and diffusion, physical factors include water purity, water depth and size of surface. It is often difficult to assess the relative importance of each factor. (Perry et al., 1984, Ward and Robinson, 1990).

3 Evaporation in Hydrology

Evaporation is an important element of the hydrological cycle. About 70% of the annual precipitation on the land surface of the earth is returned to the atmosphere by evaporation and transpiration. This is part of the source of precipitation itself in addition to the evaporation from seas and oceans, from falling precipitation and vegetation surfaces.

Hydrologists estimate reservoir evaporation by a number of methods including experimental measurement from evaporation pans, empirical formulae based on data collected from shallow lakes, water budget determination, energy budget determination, and aerodynamic determination. These methods do not yield a unique estimate (neither do the different equations based on the same method as will be shown in a table in the following sections of this article). Among them, the latter method, which is based on Dalton's equation, has been adopted as Menzel's equation and chart in concrete technology and is being widely used.

Pan and shallow lake evaporation data are conveniently published as nomographs by the U.S. National Weather Service (Kohler et al., 1955). Estimates of evaporation potentials of reservoirs as a function of solar radiation, air temperature, dew point and wind movement can be obtained from these nomographs. It should be noted that water temperature is not included as a parameter in these nomographs. Large sizes of lakes as compared to pans account for the substantially lower evaporation rates; as air moves over the lake it picks up moisture and its saturation deficit is continuously reduced as it moves further. Therefore, average evaporation per unit surface area is small as compared to evaporation from pans (Ward and Robinson, 1990). Hydrologists are interested in evaporation potential of lakes but concrete technologists should find pan evaporation data more relevant.

The importance of water temperature relative to air temperature is also very relevant in concrete technology. If the water (on surface of concrete) is warmer than air, it will have a higher vapor pressure than the saturation value at the air temperature and the vapor pressure gradient encourages intense evaporation. This is the reason why mist forms over lakes and swamps during cool calm nights (Ward and Robinson, 1990).

4 Evaporation in Drying

The rate of evaporation from a surface drying due to heat transfer solely by convection can be estimated from heat or mass transfer equations provided that the heat or mass transfer coefficient is determined (Perry et al., 1984). This was elaborated earlier by the authors, equations were given and interpretations for concrete technologists were discussed (Cebeci and Saatci, 1985). When other heat effects are involved, such as solar radiation, heat of hydration of cement in concrete, a greater evaporation rate is observed (ACI, 1991). Absorptivity of the concrete surface must be determined in order to calculate the heat gain from solar radiation. Heat of hydration should not be added directly either, because fresh concrete placed at construction site is normally not insulated but allowed to lose heat to the base or through confining formwork. Hence, concrete surfaces in shade or exposed to sunlight, fresh and hardened concrete possess different evaporation potentials under identical atmospheric conditions (temperature, humidity and wind).

5 Evaporation in Concrete Technology

The preceding discussions were not commonly encountered in concrete technology literature until 1985, whereas Menzel's equation and nomograph have been published and republished by concrete institutions of international reputation (ACI, 1991; C&CA, 1980; FIP, 1986; NSGA-NRMCA, 1960; PCA, 1966). Such wide and authoritative dissemination encouraged universal acceptance and use of Menzel's work. However, during the last decade concrete scientists and technologists have questioned the indiscriminate application of the evaporation chart (Berhane, 1984; Cebeci and Saatci, 1985; Mather, 1985; Shaeles and Hover, 1988 and 1989; Turton, 1989; Unpublished Communications, 1988). The results of different studies will be discussed and the article will be concluded with a review of the criticisms and recommendations.

Table 1 shows the rate of evaporation estimated by different equations based on Dalton's law (taken from Mather (1985) or from Table 11-1 of Veihmeyer (1964)), and by the heat transfer equation (given in Cebeci and Saatci (1985) or Perry et al (1985)). Substantially different predictions clearly demonstrate the significant influence of meteorological and physical factors under identical climatic conditions (temperature, humidity and wind).

Menzel's estimations agree with Kirkpich and Williams' pan evaporation results within $200g/m^2$-hr differences. Fitzgerald equation predicts the highest rates, up to

Table 1. Rate of evaporation estimated by different equations

			Rate of evaporation (g/m^2-hr)							
	Rel hum	Wind speed	Water/air temperature (^{0}C)							
Equation	%	km/hr	10/10	20/40	20/30	20/20	30/40	30/30	40/40	40/30
Menzel	95	5	10	–	–	20	–	36	63	571
	95	35	44	–	–	85	–	153	267	2423
	20	5	167	147	254	319	471	578	1006	1112
	20	35	710	623	1077	1353	2002	2456	4272	4726
Kirkpich	95	5	21	–	–	37	–	62	101	703
and	95	35	77	–	–	136	–	230	375	2612
Williams	20	5	239	213	344	421	594	711	1158	1266
	20	35	887	790	1279	1563	2208	2643	4301	4701
Fitz-	95	5	19	–	–	37	–	67	116	1057
gerald	95	35	90	–	–	171	–	311	541	4910
	20	5	310	272	470	590	873	1071	1864	2062
	20	35	1438	1261	2181	2740	4055	4975	8653	9573
Meyer	95	5	10	–	–	19	–	34	59	535
	95	35	12	–	–	22	–	40	70	632
	20	5	157	138	238	299	442	542	944	1044
	20	35	185	162	281	353	522	641	1114	1233
Lake	95	5	4	–	–	7	–	12	21	193
Hefner	95	35	25	–	–	47	–	86	149	1353
	20	5	57	50	86	108	160	196	341	377
	20	35	396	348	601	755	1118	1371	2385	2639
Lake	95	5	3	–	–	6	–	12	20	236
Mead	95	35	24	–	–	45	–	81	142	1650
	20	5	54	20	58	102	109	186	324	460
	20	35	377	143	409	717	760	1302	2265	3217
Heat	95	5	25			26		27	28	
transfer	95	35	117			123		129	133	
	20	5	704			743		780	816	
	20	35	3341			3523		3699	3868	

about 4000g/m^2-hr greater than Menzel's, whereas Meyer equation predicts up to 3500g/m^2-hr lower rates than Menzel's. Except for very dry, windy and warm conditions, evaporation rates from lakes Hefner and Mead agree reasonably well with each other, and as expected, they predict the lowest rates; even lower than Meyer's.

Table 1 also verifies the intense evaporation warning when the water is warmer than air. At a given water surface temperature the rate of evaporation increases as the temperature of air decreases.

It is interesting to note also that the rate of evaporation from water at 10°C and air at 10°C is greater than the rate from water at 20°C and air at 40°C. It should be apparent that evaporation is not a problem specific to hot weather construction sites; evaporation may be a serious threat to freshly placed concrete left unprotected in temperate or cool climate.

6 Conclusions and Recommendations

Menzel's formula is not a fundamental law of evaporation from concrete surfaces. It is an adaptation of Dalton's law by Menzel, and therefore, the assumptions, simplifications and limitations of Dalton's law hold for Menzel's formula as well. Moreover, the origin of his constants is not apparent from the literature (Unpublished Communications, 1988). Nevertheless, it predicts the evaporation potential from concrete surfaces covered with some underestimation as compared to hydrologists' evaporation pan observations.

Bleeding of concrete and the rate at which bleed water is made available at a concrete surface are not estimated or accounted by Menzel's formula. Concrete technologists must use their own judgement regarding the surface condition of freshly placed concrete and avoid using the evaporation nomograph for unsaturated concrete surfaces.

The recommendation in the evaporation graph that evaporation rates greater than $500g/m^2$-hr (or twice as much in some guides) are likely to necessitate precautions against premature drying needs to be elaborated. This is not a fundamental law either but a warning based on observations prior to 1960! Today, concrete made with normal and high-range water reducing admixtures contain much less initial water for bleeding. Moreover, modern fine cements of high tricalcium silicate content hydrate more rapidly and, thereby consume mixing water more rapidly as compared to mixes of 1950's. Therefore, contemporary concrete technologists must appreciate that modern mixes may be subject to plastic shrinkage cracking under less drastic drying conditions.

It should be emphasized that collection of more accurate and relevant data on the rate of evaporation from concrete in hot climatic conditions and its effect on plastic shrinkage cracking is necessary so that hot weather concreting procedures and precautions can be refined.

7 References

ACI 305R-91 (1991) **Hot Weather Concreting.** American Concrete
   Institute, Detroit.
Berhane, Z. (1984) Evaporation of water from fresh concrete and
   mortar at different environmental conditions. **ACI Journal**, 81,
   560-565.
C&CA (1980) Construction Guide: **Concreting in Hot Weather.**
   Shirley, D.E., Cement and Concrete Association Publication
   45.013, Wexham Springs, U.K.
Cebeci, O.Z. and Saatci, A.M. (1985) Discussion of Berhane (1984).
   ACI Journal, 82, 930-931.
FIP Guide to Good Practice (1986) **Concrete Construction in Hot
   Weather.** Thomas Telford Ltd., London.
Kohler, M.A., Nordenson, T.J. and Fox, W.E. (1955) **Evaporation
   from pans and lakes.** U.S. Weather Bur. Res. Pap. 38.
Mather, B. (1985) Discussion of Berhane (1984). **ACI Journal**, 82,
   931-932.
NSGA-NRMCA (1960) **Plastic Shrinkage Cracking.** Engineering
   Information Bulletin, National Sand and Gravel Association and
   National Ready Mixed Concrete Association, Washington, D.C.
PCA (1966) **Hot Weather Concreting.** Portland Cement Association,
   Concrete Information Sheet IS 14.02T, Skokie.
Perry, R.H., Green, D.W. and Maloney, J.O. (1985) **Perry's Chemical
   Engineer's Handbook,** 6th Ed., McGraw-Hill Book Co., New York.
Shaeles, C.A. and Hover, K.C. (1988) Influence of mix proportions
   and construction operations on plastic shrinkage cracking in
   thin slabs. **ACI Materials Journal**, 85, 495-504.
Shaeles, C.A. and Hover, K.C. (1989) Reply to Turton (1989). **ACI
   Materials Journal**, 86, 532.
Turton, C.D. (1989) Discussion of Shaeles and Hover (1988). **ACI
   Materials Journal**, 86, 531.
Unpublished Communications (1988) O.Z. Cebeci, R.D. Gaynor, W.E.
   Kunze, B. Mather and R.J. Van Epps.
Veihmeyer, F.J. (1964) Evapotranspiration, in **Handbook of Applied
   Hydrology.** (ed. Ven Te Chow) McGraw-Hill Book Co., New York.
Ward, R.C. and Robinson, M. (1990) **Principles of Hydrology,** 3rd
   Ed., McGraw-Hill Book Co., New York.

4   EXPERIMENTAL STUDY OF THE EFFECT OF AMBIENT CONDITIONS ON TEMPERATURES INSIDE MORTAR SPECIMENS

Y. MATSUFUJI, T. OHKUBO and T. KOYAMA
Kyushu University, Fukuoka, Japan

1.Introduction

There is a general view that concrete mixed and/or placed at a high ambient temperature, namely in a hot weather environment, may be deficient is several aspects when compared to concretes processed at normal temperature. These deficiencies include decreased workability, reduced long term strength and increased crack occurrence[1].

A recent report on the hot weather concrete [2] indicates that as far as physical properties are concern there is no drawback to the strength development of the body because of mix proportion for summer season except the degradation of the surface layer, and initial cracking in particular, may create serious problems.

From this point of view and particularly concerning the internal temperature condition of specimens, we have conducted the present work in order to evaluate the effect of various ambient conditions on the initial cracking that grows on thin concrete slabs for floor, wall and the like, which are produced in hot weather environments.

Temperature difference inside concrete body could be a principal cause of crack occurrence as it enhances stress by accelerating drying shrinkage, bleeding, etc. It is important to quantitatively evaluate the relation between ambient temperature and crack occurrence, reflecting the fact that the hot weather concrete intrinsically relates to high ambient temperatures.

The focus of this study was to evaluate the effect of ambient conditions on the temperature conditions of specimens. This in turn, is used to study the relationship between high temperature ambience and crack occurrence.

2. Outline of Experiment

2.1 Specimens
The materials of the specimens for the present experiment and the mix proportion are shown in Tables 1 and 2, respectively. Normal portland cement and beach sand are used for all the specimens with a water cement ratio of 0.5.

2.2 Conditions of Ambient Air
The relationship between the ambient conditions and specimen temperature were experimentally determined. As shown in Table 3, this study discusses how ambient temperature levels, time dependent changes of ambient temperatures, ambient humidities and wind speeds affect to the internal temperature conditions inside mortar specimens.

Concrete in Hot Climates. Edited by M. J. Walker. © RILEM
Published by E & F N Spon, 2 - 6 Boundary Row, London SE1 8HN. ISBN 0 419 18090 7.

The ambient conditions for the experiment were accomplished in a curing room of 3.0 × 3.5 × 2.4 m in size under the control of a warm and wet ambience simulation system.

Table 1　Physical properties of materials

Materials	Used Materials	Specific Gravity	Absorbed Water Ratio , %
Cement	Ordinary Portland Cement	3.15	—
Fine Aggregate	Beach Sand	2.62 (surface-dry)	2.05

Table 2　Mix proportion of mortar specimen and temperature of materials

Materials	Weight kg/m³	Temperature of Materials ,°C		
		Series A	Series B	Series C
Water	289	20	20	20
Cement	579	20	40	50
Fine Aggregate	1331	20	30	40

Series A :Planned end-of-mixing temperature = 20°C
Series B :Planned end-of-mixing temperature = 30°C
Series C :Planned end-of-mixing temperature = 40°C

Table3　Measurement Items

Experiment Series		Environmental Conditions			
Series	Item for discussing the effect	Ambient temp. ,°C	Ambient humidity ,%	Wind speed ,m/sec	End-of-mixing temp. ,°C
A	The height of Ambient temperature	15,25,35	70	0	20,30,40
B	Time-dependent change of Ambient temperature	2 Conditions (Cf. Fig*)	70	0	20,30,40
C	The height of Ambient humidity	25,35	50,70,90	0	30
D	Wind Speed	25,35	70	0, 1.0 2.0, 3.0	30

2.3 Measurement Items

The change in temperature and depth-wise distribution of temperature of mortar specimens placed under the conditions shown in Table 3 were monitored for a 24 hour period from the time of placimg.

It is thought that the temperature change of specimens after placing is affected by (1)external causes such as ambient

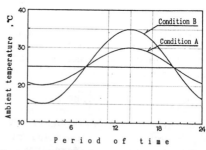

Fig.* Time-dependent changes of ambient temperature under the experiment series B

temperature, humidity and wind speed and internal causes such as (2)cement hydration heat (gain) and (3)vaporization heat due to water vaporizing from the surface (loss). From this point of view, in addition to the aforesaid temperature measurements, the heat generation rate in the specimens due to hydration and the water vaporizing rate from the surface (water losing rate) were recorded so that the time-based profiles of these rates could be plotted. All these measurements constitute the basis for ultimately evaluating how much effect such internal causes have brought onto the temperature conditions of the specimens.

(a) Temperature measurement at points distributed inside the specimens

Each specimen was a rectangular solid of 30 cm long, 30 cm wide and 10 cm high, with only the top face (hereafter "Open Face") of a 30 cm × 30 cm rectangle exposed to the ambient air and the rest of the faces isolated from the air by means of a steel sheet form which conducted only heat to/from the ambient air. In other words, drying process was allowed only through the Open Face.

The temperatures at points distributed inside the specimens and their changes with time were measured by means of copper-constantan thermocouples at every five minutes for 24 hours after placing.

The measuring points are shown in Fig.1.

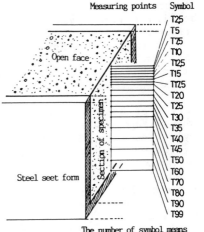

The number of symbol means the distance from open face , mm

Fig.1 Measuring points of temperature in the specimen

(b) Measurement of heat generation rate due to hydration

The hydration heat was measured for 24 hours after the end-of-mixing time by means of a micro calorimeter. The specimens were prepared with the mortar in the proportion as shown in Table 2. The measurement was made at ambient temperatures of 15°C, 25 °C and 35°C, the same as those for the principal experiment.

(c) Measurement of the amount of vaporized water (lossed water)

For measurement of the amount of vaporized water, specimens of 10 cm high, the same height as that of the specimens for the principal experiment but with a square top face (Open Face) of 5 cm× 5 cm were used for single face drying. The amount of water vaporized from the Open Face was measured at a precision of 0.01 g every 15 minutes for 24 hours. The changed weight with time was plotted to evaluate how much the temperature distribution in the specimens was affected by the heat loss due to water vaporization.

3. Test Results and Discussion

3.1 Curves of Heat Generation Rate due to Hydration

The curves of heat generation rates due to hydration in the present specimens at ambient temperatures of 15°C, 25°C and 35°C, respectively are shown in Fig. 2. These curves represent the amount of generated heat per cement weight (g) in specimens for temperature measurement.

The higher the ambient temperature, the greater the maximum value of the heat generation rate due to hydration and the quicker this value was reached.

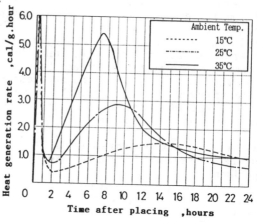

Fig.2 The curves of heat generation rate due to hydration

3.2 Curves of Rate of Heat Loss due to Vaporization

Typifying the rate of heat loss per mortar weight (g) due to vaporization from the specimens, measurements of these with the specimens having the end-of-mixing temperatures of 20°C, 30°C, and 40°C at the ambient temperature of 25°C are shown in Fig. 3. The rate of heat loss exhibited a maximum value immediately after placing and then decreased quickly for 1 to 2 hours before taking a gentle down ward slope. The higher the end-of-mixing temperature of specimen, the greater the heat loss, where the ambient temperature was consistent, within a period of 1 to 2 hours after placing.

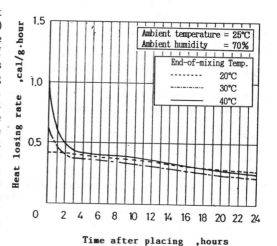

Fig.3 The curves of rate of heat loss due to vaporization

3.3 Time-dependent Temperature Change of Specimens

3.3.1 Effect of Ambient Temperature

The time-dependent temperature changes of the specimens under the respective ambient conditions shown in Experiment Series A of Table 3 were plotted in Figs. 4 which correspond to the planned end-of-mixing temperatures of 30°C. Each set of curves in these figures

36

consisted of the six measuring points of T2.5, T10, T20, T30, T40 and T50 as shown in Fig.2.
The number of symbol means the distance (mm) from Open Face.

These curves of the internal temperatures of specimens in this figure represented a general trend that they came close to the ambient temperature quickly during the period of about 3 hours after placing and then went up significantly higher than this. The profiles of temperature change observed at Point T2.5 that is closest to Open Face are schematically shown in Fig. 5 which suggest that the three principal factors listed in the previous section be contributory to this general trend.

During a period of about 3 hours after placing, the specimen temperature rises with an end-of-mixing temperature that is lower than the ambient temperature and falls with the reversed relation of the two temperatures. The difference in the shape of the profiles is more distinct with a larger difference between the two temperatures. This time

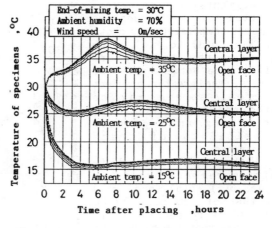

Fig.4 Time-dependent temperature changes of the specimens (Effect of ambient temperature)

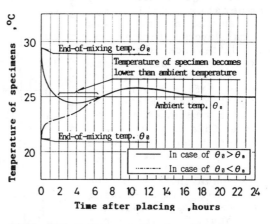

Fig.5 Typical figure of time-dependent changes of specimens (Effect of water vaporizing)

section represents the dormant period for hydration and it is thus deduced that the steep temperature changes were dominated by those due to heat conduction to/from the ambience.

At the onset of accelerated hydration , the specimen temperature started to increase, resulting in the profile having a swelling above the ambient temperature. As typified by the curves in Fig. 5, the temperatures inside specimens reached their peaks range from 7 to 16 hours after placing, though their heights varied significantly depending on the ambient temperature. It was found that the higher the ambient temperature was, the higher these peaks and the quicker the peaks arrived.

The time-dependent changes of the temperature difference between the outermost and central layers placed under condition of

Experimental Series A were plotted for the respective ambient temperatures as shown in Fig. 6(a) to 6(c). The greater the temperature difference between the two layers, the greater gradient thermal strain may be produced in a specimen, resulting in a greater possibility of crack occurrence.

With a consistent end-of-mixing temperature as in Figs. 6(a), 6(b) and 6(c) at that of 20 °C, 30 °C and 40 °C,respectively, it was found that the higher the ambient temperatures , the greater the initial temperature difference between the outermost and central layers. Furthermore, this difference exhibited a maximum value at the same time when the temperature change profile had reached its peak and the higher the ambient temperature, the greater this maximum value. Similar tendencies were observed with other end-of-mixing temperatures.

As shown in Figs. 6, the effect of end-of-mixing temperature was observed only for about 5 hours after placing as far as this temperature difference was confined in the range of the present experiment. After such points of time the specimen temperatures changed in similar profiles under different ambient conditions regardless of the end-of-mixing temperature. This brought it to light that the end-of-mixing temperature hardly affected hydration heat generation during the period of accelerated hydration, while the ambient

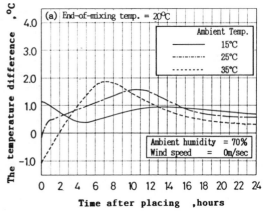

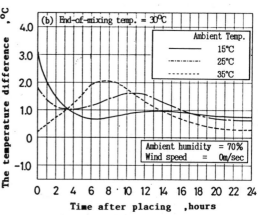

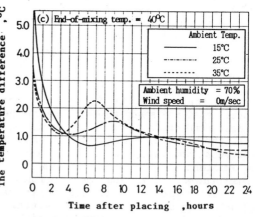

Fig.6 Time-dependent changes of the temperature difference between the outermost and central layers (Effect of ambient temperature)

temperature dictated the height of the peak of the specimen temperature profile as well as the length of time required by the peak to appear after mortar placing.

3.3.2 Effect of time-dependent change of ambient temperature

In Experiment Series B, the ambient temperature were changed with time. The effect of this tome-dependent change in ambient temperature on the difference between outer-most and central layers are shown in Fig.7. These curves represent the temperature difference of mortar specimens placed at 14:00. As shown in Fig.7, the greatest change in temperature difference was found under condition B, when ambient temperature were severely changed. Because the ambient temperature of the Condition B severely decreased when the temperature of specimen increased , consequently the temperature at the outer most layer was lower than that at the central layer.

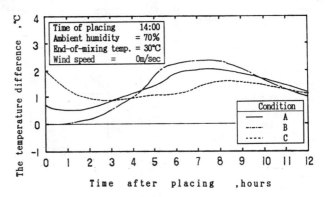

Fig.7 Time-dependent changes of the temperature difference between the outermost and central layers (Effect of environmental condition)

The effect of placing time on the temperature difference between the outer most and central layers were shown in Fig.8. As shown in Fig.8 the placing time significantly affected the temperature difference. In the range of the present experiment, the temperature difference of specimen placed at 17:00 was greatest.

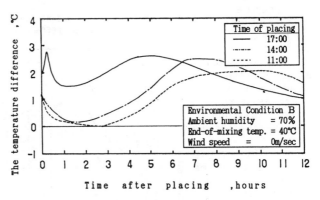

Fig.8 Time-dependent changes of the temperature difference between the outermost and central layers (Effect of time of placing)

Thus it was clarified that rapid decrease of the ambient temperature after placing was likely to cause the initial crack , increasing the temperature difference .

3.3.3 Effect of ambient humidity

The time dependent temperature changes of the mortar specimens under the respective ambient humidities in Experiment Series C of Table 2 were plotted in Figs.9(a),9(b).

These curves represented a general trend that they come close to the ambient temperature almost immediately after placing, and then started to increased as affected by heat generation due to hydration, and reached their peaks in a range from 6 to 8 hours after placing.

As shown in Fig.9 ,the higher the ambient humidity, the greater the temperature of specimen. In addition the higher the ambient humidity was ,the lesser the amount of vaporized water. As a result the heat loss due to vaporization became lower.

The effect of ambient humidities on the time-dependent change of temperature difference between the outer most and central layers are shown in Fig.10. As shown in Fig.10 the lower the ambient humidity ,

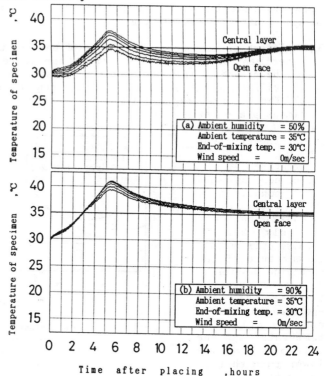

Fig.9   Time-dependent temperature changes of the specimens (Effect of ambient humidity)

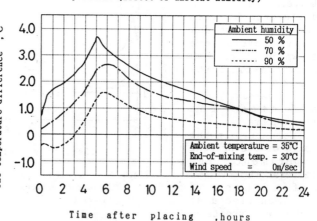

Fig.10 Time-dependent changes of the temperature difference between the outermost and central layers (Effect of ambient humidity)

the greater the temperature difference was under the same ambient temperature. It meant that the stress due to temperature difference became greater when the ambient humidity became lower.

3.3.4 Effect of wind speed

Time-dependent changes of temperature of specimen under the Experiment Series D in Table 3 are exemplified by Fig.11 which represent the results of experiment under ambient temperature of 35 and wind speed of 1m/sec and 3m/sec respectively.

Time dependent changes of temperature difference between outer most and central layers was shown in Fig.12 which represent the results of experiment under ambient temperature 35° . As shown in Fig.12 , the greater the wind speed, the greater the temperature difference during about 6 hours after placing. The effect of wind speed on temperature difference of the specimen decreased with the passage of time.

5. Conclusion

In this study, the crack occurrence in the early time in the concrete placed in hot weather environment was discussed from the viewpoint of the temperature distribution in the specimens.

The principal results obtained from this study are summarized as follows:
1. The ambient temperature significantly affects the time-dependent temperature change of the specimens,

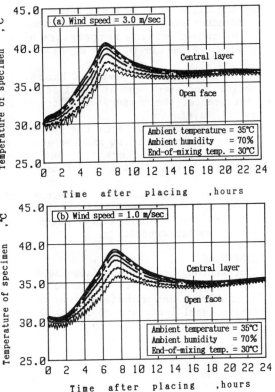

Fig.11 Time-dependent temperature changes of the specimens (Effect of wind speed)

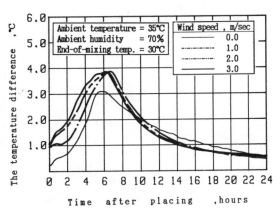

Fig.12 Time-dependent changes of the temperature difference between the outermost and central layers (Effect of wind speed)

and the higher the ambient temperature is, the more striking is the temperature profile of the specimens in the early time after placing.
2. The end-of-mixing temperature affects the internal temperature conditions inside mortar specimens only for about 5 hours after placing.
3. The effect of the ambient humidity on the internal temperature condition of specimens is significant , and the lower the humidity is , the greater the temperature difference between the outer most and central layers is.
4. The wind speed affects the internal temperature conditions for a few hours after placing and the greater the wind speed, the greater the temperature difference. But the effect of wind speed slowly decreases with the passage of time.

This study examined the internal temperature of mortar specimens through a 24 hours period after placing . The results are expected to be helpful in evaluating preventive measures against crack occurrence in hot weather concreting.

REFERENCES
1]Hot Weather Concreting, Japanese.Architecture Standard Specification (JASS 5), pp287-297 (in Japanese)
2] Problems and countermeasures of concreting in hot weather, Architectural Institute of Japan, October 1989 (in Japanese)

5 RECHERCHE SUR BÉTON AVEC GRANULAT LATERIQUE

(Research on concrete with lateritic aggregate)

N. P. BARBOSA
Technology Centre, Federal University of Paraiba,
João Pessoa, Brazil
M. B. CHAGAS
Science and Technology Centre, Federal University
of Paraiba, Campina Grande, Brazil

Abstract
In this paper some results of research about lateritic
stones as concrete aggregate are presented. Properties of
lateritic stones are shown. Mechanical characteristics of
lateritic concrete are given. Experimental data obtained
with beams made from reinforced lateritic concrete are
presented and discussed.
Keywords: Laterite, Lateritic Concrete, Reinforced
Lateritic Concrete Beams.

1 Introduction

Dans la région tropical de la Terre on trouve souvent des
sols de coloration rouge, appelés sols latéritiques. Ce nom
vient du latin "later" que signifie brique, parce que ces
sols, quand soumis à des procés de mouillage et secheresse,
concrétionent, et, en devenant durs, peuvent etre coupés
sous la forme de briques.

Dans des grandes parties du territoire brésilien, comme
la region amazonienne, les concrétions latéritiques sont
trés abbondantes. Par contre, les roches d'origine ignée,
avec lesquelles on prepare normalment le béton au Brésil,
sont trés rares en plusieurs parties de cette region. Le
transport des granulats fait donc s'accroitre le prix du
béton.

En tenant compte de la necessité d'étudier les matériaux
locaux pour l'emploi dans la genie civile, de sort à faire
baisser le prix des travaux, à l'Université Federale de
Paraiba on développe une ligne de recherche sur les sols
latéritiques, y compris l'emploi des concrétions
latéritiques dans le béton et son usage en pièces armées.
Le béton avec granulat latéritique est ici appellé béton
latéritique (B.L.).

À la suite, on présente quelques résultats des études
faits jusqu'à maintenant. On montre certaines propriétés
des pierres latéritiques et du béton fait avec elles. Les
donnés experimentaux obtenus par l'essai de poutres armés
sont aussi indiqués et comentés.

Concrete in Hot Climates. Edited by M. J. Walker. © RILEM
Published by E & F N Spon, 2 - 6 Boundary Row, London SE1 8HN. ISBN 0 419 18090 7.

2 Travaux précédents

En raison d'être très abbondants sur la surface de la Terre les sols latéritiques ont été et continuent à être étudiés par des nombreux chercherurs. Dans les années soixante, l'UNESCO(United Nations Educational Scientific and Cultural Organisation) a montré un considérable intérêt sur le matériau latéritique et publie une première révision sur les recherches à ce sujet, Magnien (1966). Application de la latérite dans le béton on n'y trouvait pas encore.

Au debut des années soixante dix, Thomas e al. (1971) commencent à exploiter les pierres latéritiques pour la fabrication du béton.

En suite, au Nigeria, Adepegba (1975) (1977) et Balogun et al. (1982) ont developpé des recherches sur le remplacement du sable par materiau latéritique pour obtenir le "béton laterisé".

Madu (1979) a étudié les pierres latéritiques de sept gisements au Nigeria et sont emploi comme granulat dans le béton.

Au Brésil, Azevedo (1983) a étudié la viabilité d'utilisation des pierres latéritiques de l'Etat de l'Acre, dans la région amazonienne, pour fabbriquer de B.L.

Souza et al. (1985) ont présenté une étude de l'influence du temps de lavage des agregats latéritiques (dans la betoniere) sur la résistance du béton.

Encore sur le B.L. d'autres travaux ont été faits à l'Université Federale de Paraiba par Pompeu Neto (1976), Carvalho (1981), Carvalho (1984), Chagas Filho (1986), Costa et al. (1987) etre autres.

Les recherches montrent que avec certaines concrétions latéritiques on peut obtenir des bétons de résistance moyenne (25-30 MPa) et avec d'autres on ne réussit qu'à avoir des bétons de faible résistance à la compression (15-18 MPa). Ces résistances sont compatibles avec la realité locale et on a donc commencé à étudier le comportement de pièces structurales armées en béton latéritique. Avant de présenter les resultats des essais expérimentaux sur poutres, on comente sur les pierres latéritiques et sur le béton fabriqué avec elles.

3 Caractéristiques des granulats latéritiques

Les proprietés chimiques, physiques et mécaniques des granulats latéritiques changent selon le gisement d'origine du material. C'est pourquoi on a coleté des echantillons de sols et des concrétions de plusieurs gisements au Nord-Nordest brésilien.

Une des plus particulieres caractéristiques des pierres latéritiques c'est le haut taux de sesquioxides de fer et de alumine (Fe_2O_3 et Al_2O_3). Ensemble au silice sous la forme de SiO_2 constituent presque la totalité du materiau.

Tableau 1. Composition des pierres latéritiques

mineral	gisement			
	1	2 (%)	3	4
Si O_2	24	41	31	24
Fe_2O_3	40	38	47	53
Al_2O_3	23	13	13	15

On trouve encore des petites pourcentages des oxides de Titanium, Calcium, Magnesium, Sodium et Potassium. Le tableau 1 donne une idée des pourcentages des principaux composants de pierres latéritiques de quatre gisements divers (1-Sapé, 2-Mosqueiro, 3-São Luiz, 4-Rio Branco).

Quant aux caractéristiques physiques, une des plus notables différences par rapport aux granulats ignées est l'absorption. Aprés une heure d'imersion, les granulats latéritiques peuvent absorver de 5 a 10% de son poids en eau. Cela est dû à son coeficient de vide elevé.

La masse spécifique des granulats latéritiques varie entre 2.40 et 2.80 kg/dm . Elle est proche de celle des pierres granitiques utilisées au Nord-est brésilien (2.80 kg/dm).

Une autre caractéristique des agregats latéritiques c'est qu'ils présentent, dans l'état naturel, un haut taux de materiaux pulvérulents. Un simple lavage peut reduire ce taux a des valeurs compatibles avec celles désirables pour la fabrication du béton.

Les proprietés mécaniques des concrétions latéritiques sont toujours inférieures à celles du materiau ignée. Ce fait devait être attendu, vu que les derniers sont originés sous des hautes pressions au centre de la Terre, au pas que les pierres latéritiques se forment sous la pression atmosphérique. L'essai d'abrasion Los Angeles (tableau 2) montre la plus grande fragilité du materiau latéritique.

La durabilité des pierres latéritiques dans de milieu alcalin du béton mérite encore des études plus aprofondies. Cependant, on a déjà fait, avec deux graviers latéritiques, des essais sur la reativité de ces agregats, selon le critère de Mielenz et Witte (1948). L'un des granulats

Tableau 2. Abrasion Los Angeles des pierres latéritiques

gisement	1	2	3	4	granit
Los Angeles (%)	42	65	63	41	13

s'est montré satisfaisant et l'autre, très proche de la valeur recommendée (Carvalho, 1984). En plus, l'examen visuel du béton de pièces essaiées en Laboratoire et aprés exposées à la pluie et au soleil, pour déjà plus de sept ans n'indique aucun problème avec les graviers latéritiques.

4 Caractéristiques du béton latéritique

La masse spécifique, comme de reste les autres proprietés physiques et mécaniques du B.L. dependent du gisement d'origine du gravier et de la composition du béton. Dans de nombreuses mesures, on a trouvé des valeurs entre 2.23 et 2.35 kg/m .

L'accroissement de la résistance avec l'age du béton latéritique est semblabe à celle du béton normale.

Le même se peut dire de la variation de la résistance avec la dosage en eau.

Quant'à l'augmentation de la résistance du B.L. avec le contenu en ciment, à cause de la plus grande fragilité des pierres latéritiques, elle a une limite, comme montre la courbe de la Fig. 1 (Azevedo, 1983). Ce graphic indique qu'à partir d'un certain point, il ne vaut pas la peine d'augmenter la quantité de ciment pour obtenir un béton plus résistant. La rupture a lieu par le fendillement des graviers latéritiques.

Le comportement du béton latéritique soumis a des efforts uniaxiales de compression est semblabe a celui du béton de granulat granitique (B.G.). La Fig. 2 montre des courbes typiques obtenues par un compressomètre instalé à des éprouvetes cylindriques (15 cm x 30 cm).

Malgré que les courbes contrainte-deformation aient la forme semblabe, le module de Yong du B.L. est plus bas que celui du B.G. Pour l'estimer sous la forme proposée au Code Model du CEB (1991), on suggere d'utiliser le paramètre α_β

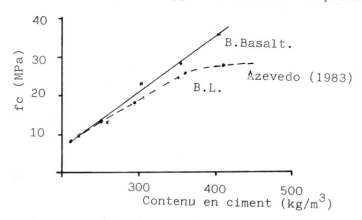

Fig.1. Courbes résistance à la compression-taux de ciment.

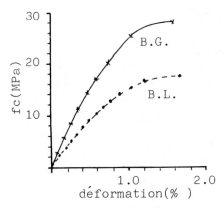

Fig.2. Curves contraite-déformation du béton.

égale a 0.60 dans l'équation (1).

$$Ec = \alpha_{\beta}\, 21500\, [fcm/10]^{1/2} \tag{1}$$

avec f_{cm} étant la résistance moyenne du béton à 28 jours.

La relation entre résistance à la traction et résistance à la compression est semblabe à celui du béton normal. La Fig. 3 montre des résultats de plusiers essais sur éprouvetes de béton latéritique obtenues par différents chercheurs. La résistance à la traction a été obtenue par l'essai de fendillement. Chaque point représente la moyenne de deux, trois ou même six éprouvetes.

On peut voir que la relation entre la résistance à la traction et la résistance à la compression est plus près de la courbe limite inferieure du CEB. Pour de B.L., une droite peut représenter bien ce rapport.

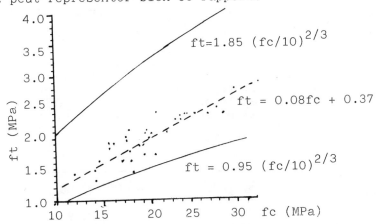

Fig.3. Relation entre résistence à la traction et résistance a la compression du B.L.

5 Expérimentation sur poutres

Pour étudier le comportement du béton latéritique en poutres structurelles, on a procédé à des essais expérimentaux de plusieurs pièces au Laboratoire de Structures de l'Université Federale de Paraiba, à Campina Grande, Brésil.

Des quatorze poutres testées (B1 à B14), dont on peut voir les caractéristiques à la Fig. 4, trois étaient en béton de granulat granitique: B5, B7, B9. On a utilisé toujours la même composition du béton: 1:2.3:2.7, avec de petites variations en la dosage en eau pour donner une affaissement de 2 à 5 cm (slump test). Le matériau latéritique était originé de plusieurs gisements brésiliens. Le sable était venu de fleuve prochain et le ciment était type pozolanique (32 MPa).

Le chargement était apliqué incrementalement par machine hydraulique Amsler. Les déformations des matériaux étaient mesurées avec des "strain gages" électriques de 100 mm (béton) et 20 mm (acier). Les flèches étaient obtenues avec extensomètres de 0.01 mm de précision.

Dans la figure 5, on peut voir les courbes charge-flèches obtenues pour plusieurs poutres. On peut noter que le comportement des pièces en béton latéritique est semblabe à celui des pièces en béton granitique, bien que les premières soient plus flexibles.

Toutes les poutres, sauf B6, B7 et B13, ont présenté rupture par flexion.

La norme brésilienne de béton armé, ABNT (1982), considère atteint l'état limite par deformations excessives lorsque la flèche maximale est égale à L/300 (L étant la distance entre centre d'apuis). Au tableau 3 on peut voir les charges d'utilisation. Fut, et ses rapports entre poutres en B.L. et en B.G. On peut noter que malgré que la résistance du B.L. soit bien inférieure à celle du B.G., les charges d'utilisation sont proches.

Selon la norme citée, pour obtenir l'armature transversale, il est permis l'emploi d'une contrainte de cisaillement reduite $\tau_d = \tau_{wd} - \tau_c$. τ_{wd} est la contrainte de cisaillement conventionelle, égale a $Vd/(b.d)$ et τ_c la parcelle de la contrainte de cisaillement qu'on peut attribuer resistée par le béton comprimé et par les armatures longitudinales, donné par: $\tau_c = 0.15\sqrt{f_{cк}}$

Tableau 3. Charges d'utilisation

poutre	B1	B2	B3	B4	B5	B6	B7	B8	B9
Fut (kN)	20.0	21.5	22.5	21.5	24.5	18.5	22.5	20.0	25.5
Fut/FutBG	0.82	0.88	0.92	0.88	1.00	0.82	1.00	0.78	1.00

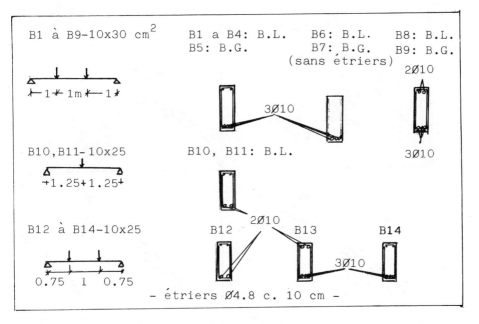

Fig.4. Caractéristiques des poutres.

En considérant la résistance caractéristique du béton, fck, égale à la résistance moyenne à j jours d'âge, fcj, moins 6.6 MPa (ABNT (1982)), étant Vmax l'effort tranchant maximal considéré comme valeur de calcul, on peut organiser le tableau 4. D'après ce tableau, on note que le rapport entre les contraintes τ_{wd} et τ_c a été plus haut pour la poutre en B.L. Cela veut dire que l'effort tranchant ultime a été relativement plus élevé pour la poutre en béton latéritique.

On a aussi fait une étude comparatif entre le moment flechisant obtenue expérimental et théoriquement. Pour cela on a developpé un programme de calcul pour obtenir le moment résistant des sections. On a utilisé le modèle proposé au CEB avec le diagramme parabole-retangle pour le béton et en tennant compte des courbes contraintes deformations des aciers brésiliens. La résistance de calcul du béton a été obtenue par fcd = fck/1.4.

Tableau 4. Contraintes de cisaillement des poutres B6 et B7

poutre	fck	Vmax	τ_{wd}	τ_c	τ_{wd}/τ_c
B6 (B.L.)	10.4	23.0	0.82	0.48	1.71
B7 (B.G.)	23.8	29.5	1.05	0.73	1.44

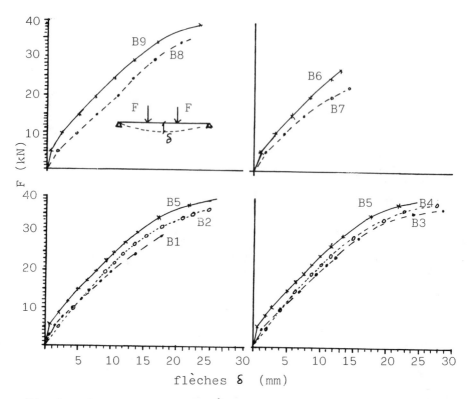

Fig.5. Courbes charge-flèche.

Avec les moments théoriques, Mt, obtenus par le programme et en considérant le moment expérimental, Me, comme celui correspondant à la charge maximale obtenue en laboratoire, on a creé le tableau 5.

La première poutre testée a eu un problème de instabilité due au systeme d'application de charges lorsque le moment était 30 KN.m, soit 1.5 fois le moment théorique.

La poutre B13, avec haut taux d' armature, a présenté rupture par cisaillement, lorsque le moment avait atteint dejá 1.7 fois le moment résistant théorique.

Avec les autres poutres, la rupture a y lieu avec charges 1.7 a 2.1 fois superieures à celles correspondantes aux moments théoriques.

Aucun problème d'adhérence a été observé entre le B.L. et les armatures.

6 Conclusions

Les caractéristiques physiques et mécaniques des granulats latéritiques varient avec le gisement d'origine.

Ces granulats sont poreux et absorbent de l'eau.

Tableau 5. Moments flechisants expérimentaux et théoriques

poutre		age (jours)	fcj (MPa)	fck (MPa)	Me (kN.m)	Mt (kN.m)	Me/Mt	Me/Mut
B1		28	15.5	8.9	–	20.5	–	–
B2		89	23.4	16.8	39.5	22.8	1.7	2.5
B3		48	18.8	12.2	42.2	19.1	2.2	3.1
B4		49	18.9	12.3	41.2	19.1	2.1	3.0
B5	(BG)	56	30.3	23.7	46.2	24.3	1.9	2.8
B8		28	12.7	6.1	47.0	24.5	1.9	2.9
B9	(BG)	29	29.3	22.7	48.6	26.0	1.9	2.9
B10		29	17.2	10.6	23.7	11.0	2.1	3.0
B11		42	17.2	10.6	23.7	11.0	2.1	3.0
B12		35	17.6	11.0	22.1	11.2	2.0	2.8
B13		37	17.6	11.0	–	–	–	–
B14		80	19.4	12.8	27.4	14.8	1.9	2.6

Bien que les pierres latéritiques soient plus fragiles que les roches ignées, avec certaines d'elles on peut obtenir de bétons de moyenne résistance et avec d'autres, de bétons de faible résistance.

Avec la même composition, la résistance à la compression du B.L. peut varier, selon le gisement, de 50 a 90% de la résistance du B.G.

Le module d'Yong du B.L. est inferieur à celui du B.G.

Le béton latéritique est legèrement moins lourd que le béton de granulat granitique.

La variation de la résistance du B.L. avec l'âge et avec le rapport eau/ciment est semblabe à celle du B.G. Par contre, l'augmentation de la résistance du B.L. avec le contenu en ciment est diverse et limité.

Le comportement en flexion des poutres en B.L. est pareil au comportement des poutres en B.G., bien que les premières soient plus flexibles.

L'évolution de la déformation du béton comprimé et de l'acier tendu (non presentés dans ce travail) sont en tout semblabes, soit la poutre en B.L., soit en B.G.

La charge d'utilisation des poutres en béton latéritique a été en moyenne près de 85% de celle des pièces en B.G.

La charge ultime expérimentale des poutres a été 2.5 à 3 fois superieure à la charge d'utilisation.

Le rapport entre le moment de rupture expérimental et les moment résistent de calcul théorique présente presque les mêmes valeurs pour les poutres en B.L. ou en B.G. La présence de l'effort tranchant n'a pas influencé sur ce rapport.

Le modèle de flexion du CEB semble pouvoir être emploié au béton latéritique avec sûreté.

7 Références

ABNT (1982) Projeto e Execução de Obras de Concreto Armado. **Assot.Brésilienne Normes Techniques**. Rio de janeiro.

Adepegba, D. (1975) The efect of water content on the compressive strength of laterized concrete. **J. Testing and Evaluation**, vol 3 n.6 nov 1975, pp 449-453.

Adepegba, D. (1977) Structural strength of short axially loaded columns of reinforced laterized concrete. **J. Testing and Evaluation**, vol.5, n.2, mar.1977 pp 134-140.

Azevedo, B.A. (1983) La laterite acreane utilise comme granulat dans le béton. (en portugais). **Meeting IBRACON 1983**, Inst. Bras. do Concreto, San Paolo pp 1-61.

Balougun, L.A. and Adepegba D. Effet of varying sand content in laterized concrete. **Int. Journal of Cement and Lightweight Concrete**, vol. 4, nov. 82.

Carvalho, J.B.Q. (1981) The use of lateritic concretions in the fabrication of concrete. **Brasilien Symposium on Tropical Sols in Engineering**, Rio de Janeiro.

Carvalho, J.B.Q. (1984) Lateritic aggregate used to fabricate concrete. **Boul. Int. Assoc. Engineering Geology**, n. 30, Paris, pp 461-464.

CEB (1991) CEB-FIP Model Code 1990. Final Draft. **Comite Euro-international du Béton** Boul.d'inform. n.203-205.

Chagas Filho, M.B. (1986) Concretions latéritiques: propiétés basiques et son emploi en poutres isostatiques en flexion simples (en portugais). Tese de Master, **Dept. Eng. Civil Univ. Fed. Paraiba**, Brésil.

Costa, C.R.V. and Lucena, F.B. (1987) Utilisation d'un sol latéritique pour la fabrication du béton de ciment portland (en portugais). **Meeting de Pavémentation**, Soc. Bras. Pavimentação, Maceió, Brésil.

Madu, M. (1980) The performance of lateritic stones as concrete aggregate and road chipyngs. **Materiaux and Constructions** vol. 13, n.78, pp 403-411.

Magnien, R. (1966) Revew of researches on laterite. **Unesco Publication**, pp 1-148.

Mielenz, R.C. and Witte L.P. (1948) Test used by bureau of reclamation for identifying reactive concrete aggregates. **Proc. ASTM 48**, pp 1071-1103.

Pompeu, P.P. (1976) Une étude sur les propriétés de résistance mécanique du béton latéritique (en portugais) Tese de Master, **Dept. Eng. Civil Univ. Fed.Paraiba**, Bres.

Souza, A.C. and Pinto, A.F. (1985) Effet du temps de lavage des granulats sur la résistance du béton latéritique (en portugais). **Pesq. no Maranhão**, Univ. Fed.Ma. n1 pp 1-8.

Thomas, K., Lisk, W.A.A. (1971) Investigation into the suitability of crushed laterite rocks for use as coarse agregate for concrete, in **Concrete and Reinforced Concrete in Hot Countries**, Haifa, vol. 1 pp.183-198.

PART TWO

PERFORMANCE

6 INFLUENCE OF INITIAL CURING TEMPERATURE ON CONCRETE PERFORMANCE IN VERY HOT ARID CLIMATES

T. SUZUKI
Shimizu Corporation, Tokyo, Japan

Abstract
In order to obtain the fundamental data for hot weather concreting, three kinds field tests were carried under the climate condition of high air temperature from 40℃ to 45℃ , low relative humidity from 18% to 22%. Series 1 was aimed to know the influence of concrete temperature on the water evaporation in concrete and the compressive stength. Difference of compressive strength were about 2 N/mm2 for initial concrete temperature from 32℃ to 35℃. Series 2 was to know the make method of concrete temperature under 32℃. The method that cold waters were scattered on aggregates under the sun shines have the effect of decrease of initial concrete temperature. Series 3 was mainly investigated that influence of concrete condition when concrete were agitated in truck agitater until 2 hours after mixing by bating plan. Three kinds of concrete with different initial temperature were agitated. Base upon the test results, the influence of initial concrete temperature and other factors on slump loss was analysed. The good quality concrete of two hundred thousand m3 were placed on one year according to the specification of concreting in very hot and arid climates were made by these test results.

Keywords: Hot weather concreting, slump loss, Initial temperature, Temperature rise, Prolonged mixing, Cold water, Retarder, Compressive strength,

1. Introduction

Most specifications for hot weather concreting require that concrete temperature as placed should not exceed a certain maximum limit as reasons to do not occur following matters.
   · Increased water demand for required slump
   · Reduced strength and durability due to increased water demand
   · Difficulties with handling, compacting and finishing due to rapid slump loss or accelerated set
This report describes the results of three kind field tests were carried under the climate condition of high air temperature from 40℃ to 45℃, low relative humidity from 18% to 22%.

Concrete in Hot Climates. Edited by M. J. Walker. © RILEM
Published by E & F N Spon, 2 - 6 Boundary Row, London SE1 8HN. ISBN 0 419 18090 7.

(1) Influence of initial concrete temperature on the water evaporation
    in concrete and the compressive strength.
(2) The method to make initial concrete temperature under 32℃.
(3) The change of concrete condition when concrete were agitated in
    truck agitater until 2 hours in 3 season.
    The good quality concrete of two hundred thousand m3 were placed
on one year according to the specification of concreting in very hot
and arid climates were made by these test results.

2. Test plan

2.1 Materials
    Following materials were used for concrete.
· Cement:
        Normal portland cement(Type Ⅰ), Kuwait cement Co., Kuwait
· Fine aggregate:
        Sand produced in Kerbala area. Specific gravity 2.60,
        Fineness modulus 2.70, Absorption 1.50%
· Coarse aggregate:
        Gravel produced in Samara area. Specific gravity 2.65,
        Fineness modulus 6.70, Absorption 0.60%
· Admixture:Water-reducing agent(Ligin type)
        Master Builders Co.,Japan

2.2 Test procedure
    In order to obtain the fundamental data for hot weather concreting,
next three kinds test were carried under the climate condition of
high air temperature from 40℃ to 45℃ and low humidity from 18% to
22%. The main points aimed at each series are as follows.

(1) Series 1
    Concrete initial temperature were controlled by adding materials of
different temperatures into the mixier. The mixier is tilted-drumed
mixier of 100 liter volume per 1 batch. After mixing of three minutes,
slump and initial concrete temperature were measured immediately, and
 then three specimens (30cm×27cm×3.8cm) for water evaporation in
concrete and three specimens (15cmφ×30cm) for compressive strength
were taken. Measure of concrete temperature and weight loss for water
evaporation in concrete were repeated at thirty minute intervals
until 150 minute or 300 minute. Specimens for compressive strength
were stored in a shed,capped at the age of one day, demoulded at the
age of 2days and then cured in field water tank under natural temperature of about 28℃ until they were tested at the age of 7days and 28
days. Specimens at the age of one day were capped by gypsum and were
tested. Mix proportion are shown in Table 1.

(2) Series 2
    To check the decrease effect of aggregates temperature were
measured temperature in different depth of aggregates stock yard
after cold waters(about17℃) were scattered on sands and gravels
 under the sun shines. The measure method of temperature in the
aggregates is shown in Figure 1.

Waters were cooled by chiller machine. The methods to get initial concrete temperature under 32°C on specifications were investigated by mixing at materials of different temperature included cold waters and cold aggregates. Mix proportion are shown in Table 1.

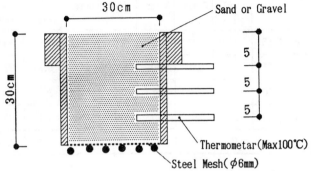

Figure-1 Temperetare test Method (Series 2)

(3)Series 3
 In this series, three kinds of concrete with different initial temperature, with and without retarder were agitated in truck mixer of nominal capacity $4m^3$ until two hours after mixing. About 50 liters of concrete sample per one measurement was discharged from the drum into concrete cart.With this sample,temperature of concrete and slump were measured immediately,and then three specimens(15 φ cm× 30cm) for compressive strength were taken.This operation was repeated every 30 minutes after mixing and continued until 120 minutes. The curing met -hods and test ages of compressive strength specimens have same in series 2. Mix proportion are shown in Table 1.

3. Test results and discussion

3.1 Sreies 1
 The change of weight loss were caused by water evaporation in con- crete are shown in Figure 2. The change of concrete temperature are shown in Figure 3.The temperatures of each materials and initial con- crete, and test results of compressive strength are shown in Table 2 and Figure 4.The concrete lowered initial temperature,the temperature raised rapidly with the lapse of time but water evaporation ratio slowed. This trend is quite natural, because the main driving force to raise the concrete temperature is the temperature difference betwe en ambient temperature and that of concrete.This compressive strength test results indicates that the higher the initial temperature, the higher the strength retention ratio to initial strength, and it sugg- ests tendency that lowering the initial temperature would have same strength by long ages curing. The higher the initial temperature, the rapider the water evaporation ratio rise,so the protect of concr- ete surface by moisture curing are necessary to prevent plastic- shrinkage cracking.

Table-1 Mix proportion of Concrete

Series	Seasons	W/C (%)	Slump (cm)	Mix proportion (kg/m³)						Max temp (℃)
				Cement	Water	Sand	Gravel	Admixture		
								Weight	Name	
1	Summer	55	15	309	170	686	1178	0.93	300R	44
2	Summer	55	15	309	170	686	1178	0.93	300R	45
3	Spring	50	15	334	167	702	1171	2.5	NO.84	18
	Autum	53	15	321	170	676	1178	2.4	300R	25
	Summer	55	15	303	167	728	1171	1.94	300R	43

Table-2 Compressive Strength Test Results (Series 1)

NO.	Temperetare(℃)					Slump (cm)	Compressive Stength (N/mm²)		
	Cement	Water	Sand	Gravel	Concrete		1Day	7Days	28Days
1	43.5	14.5	30.5	32.0	32.0	13.5	16.4	23.1	28.5
2	43.5	24.5	30.5	33.0	33.0	15.0	17.1	27.3	29.4
3	43.5	31.0	31.0	40.5	35.0	14.0	18.3	26.2	29.8

Table-3 Concrete Temperature on Material Temperature (Series 2)

NO.	Cement (℃)	Water (℃)	Sand (℃)	Gravel (℃)	Concrete (℃)
1	44.0	30.0	44.0	42.0	39.0
2	44.0	30.0	44.0	28.0	34.0
3	44.0	17.0	44.0	28.0	30.0
4	44.0	17.0	33.0	28.0	28.0
5	35.0	17.0	44.0	28.0	29.0

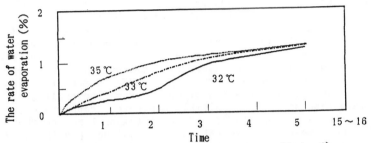

Figure-2 Water evaporetion vs time after mixing (Series 1)

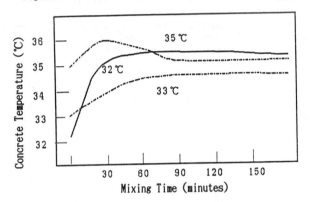

Figure-3 Concrete Temperature vs mixing time (Series 1)

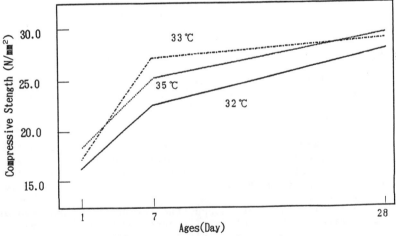

Figure-4 Compressive Strength vs Ages (Series 1)

3.2 Series 2

The change temperature of sands and gravels under the sun shines are shown in Figure 5 and Figure 6. These data indicated average of three days in same ambient temperature. The change temperature after cold waters(about 17℃) were scattered on sands and gravels when the air temperature became the maximum in day are shown in Figure 7 and Figure 8. These date indicated average of six days at same ambient temperature. Fog spraying of aggregates can reduce aggregates temperature by latent heat of evapolation of cold waters.The cooling effect is more gravels than sands. The initial concrete temperature were measured by mixing at materials of different temperature are shown in Table 3. The concrete used cold waters and cooling gravels can get initial concrete temperature about 30℃.Sands did not cooling because wetting of sands tends to cause variations in surface moisture and complicates slump control.The measured results of initial temperature of concrete that are mixed at temperatures of each materials,cements :45℃,waters:17 ℃,sands:44 ℃,gravels:28℃,ambient temperature:45℃ are shown in Figure 9. The variations of each materials temperature were ±2℃.

3.3 Series 3

Test results of three kinds are shown in Table 4. Test results on slump, temperature and compressive strength are shown in Figure 10 ～ Figure 12. This test results may indicate that the effect of lowering the concrete temperature would not be effective to minimize the slump loss under high ambient temperature. The slump loss after 30 minuts agitated on three kinds were about 5cm. Rate of temperature rise obtained from the test is nearly constant irrespective of time from mixing till about two hours,and apporoximately 2℃ /hr in average for the case initial temperature of 30℃ , and 1℃/hr for 16℃. The agit-ated times of concrete were indicated necesary under 30 minuts for get concrete temperature under 32℃ in very hot arid climate.
Concrete strength decreased as the period from mixing till placing became longer.

4. conclusion

In order to obtain the fundamental data for hot weather concreting. field tests were carried out. Problems about rapid slump loss as a focal point, temperature rise, strength and others were investigated and following conclusion were obtained.

(1) The concrete lowered initial temperature,the temperature raised rapidly with the lapse of time but water evaporation ratio slowed.
(2) This compressive strength test results indicates that the higher the initial temperature,the higher the strength retention ratio initial strength.
(3) The higher the initial temperature, the rapider the water evapora-tion ratio rise, so the protect of concrete surface by moisture curing are necessary to prevent plastic-shrinkage cracking.
(4) Fog spraying of aggregated can reduce aggregates temperature by latent heat of evapolation of cold waters.

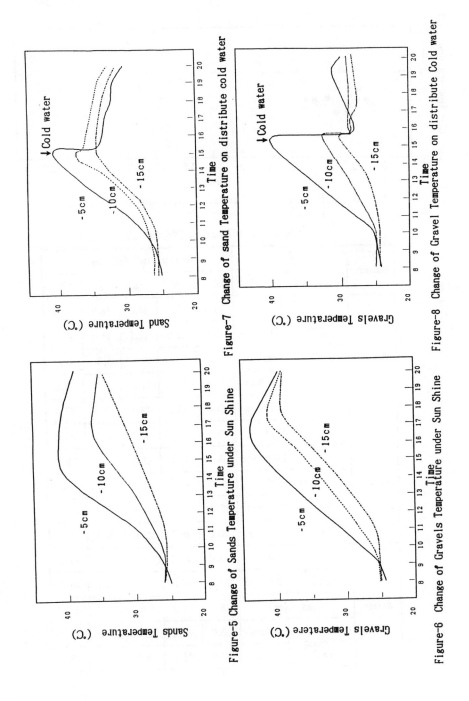

Figure-5 Change of Sands Temperature under Sun Shine

Figure-6 Change of Gravels Temperature under Sun Shine

Figure-7 Change of sand Temperature on distribute cold water

Figure-8 Change of Gravel Temperature on distribute Cold water

61

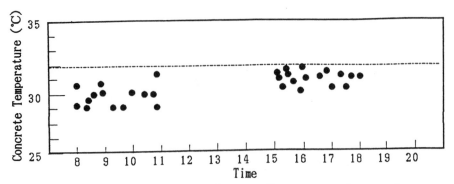

Figure-9 Test results of Concrete Temperature (Series 2)

Table-4 Test Results (Series 3)

Seasons	Items		Mixing Time (minutes)				
			0	30	60	90	120
Spring	Temperature (°C)	Air	18.0	18.0	18.0	18.0	18.0
		Concrete	16.0	17.0	17.5	18.0	18.5
	Slump (cm)		16.0	9.5	6.5	5.5	4.5
	Compressive Strength (N/mm²)	7 Days	24.5	21.4	23.8	21.8	23.8
		28Days	31.9	30.3	29.4	27.8	27.4
Autum	Temperature (°C)	Air	19.0	20.5	22.0	22.0	25.0
		Concrete	18.5	19.5	21.0	21.5	23.5
	Slump (cm)		14.5	8.5	5.5	4.5	3.5
	Compressive Strength (N/mm²)	7 Days	27.0	25.7	24.6	24.7	23.6
		28Days	30.6	28.7	26.6	26.7	36.9
Summer	Temperature (°C)	Air	41.0	41.2	43.5	41.5	41.0
		Concrete	30.0	32.0	33.0	33.0	33.5
	Slump (cm)		16.0	11.0	9.0	8.0	6.0
	Compressive Strength (N/mm²)	7 Days	25.7	25.7	22.8	24.3	22.7
		28Days	32.2	34.1	28.1	29.8	29.8

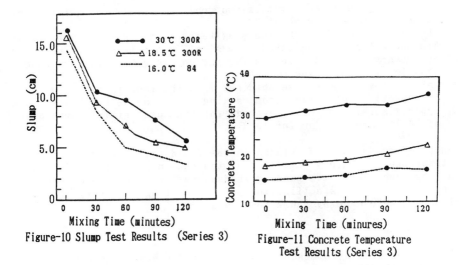

Figure-10 Slump Test Results (Series 3)

Figure-11 Concrete Temperature
Test Results (Series 3)

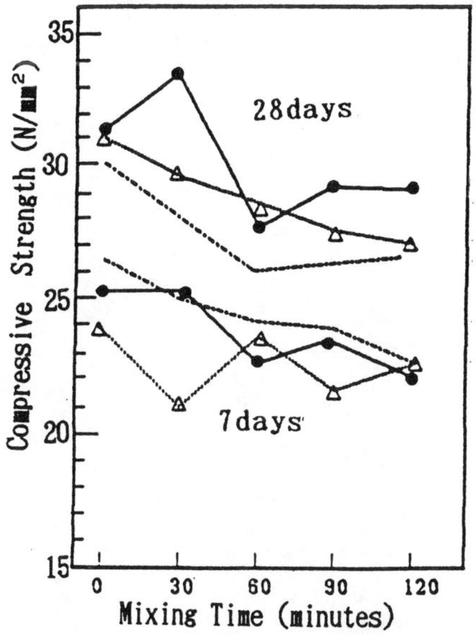

Figure-12 Compressive Strength
Test Results (Series 3)

(5) The concrete used cold waters and cooling gravels can get initial concrete temperature about 30℃.

(6) The slump loss after 30 minuts agitated on three kinds were about 5cm, and then the agitated times of concrete were indicated necceesary under 30 minutes for get concrete temperature under 32℃ in very hot aride climate.

(7) Rate of temperature rise obtained from the test is neary constant irrespective of time from mixing till about two hours, and apporoximately 2℃/hr in average for the case initial temperature of 30 ℃ ,and 1 ℃/hr for 16 ℃ .

Concrete produced and placed in the field will experience such a temperature history similar to the case of this test. The good quality concrete of two hundred thousand m^3 were placed on one year according to the specification of concreting in very hot arid climates were made by these test results.

5. References

(1) ACI Standard, Recommended Practice for Hot Weather Concreteing (ACI 305-91)

(2) Rahel Shalon, 'Concrete and Reinforced Concrete in Hot Countries' Materiaux et Constructions, Vol.11, No.62

(3) 'Guide for Use of Hot Weather Concreting' , Japan Architectural Institute, July, 1992.

7 INFLUENCE OF TEMPERATURE, DRYING AND CURING METHOD ON PROPERTIES OF HIGH STRENGTH CONCRETE

M. A. SAMARAI
College of Engineering, University of Baghdad, Iraq
K. F. SARSAM
University of Baghdad, Iraq
N. M. KAMALUDDEEN
College of Engineering, University of Baghdad, Iraq
M. AL-KHAFAGI
National Centre for Construction Laboratories, Baghdad, Iraq

Abstract
In this work the influence of mix temperature and/or relative humidity (RH) is studied for HSC. This is the second paper in a programme intended to study HSC behaviour in hot and dry environment (Samarai et al., 1987). Research indicates similarities as well as differences in the influence of environment on HSC and NSC.
Mix temperature rise is found to be harmful to both tensile and compressive strengths of concrete, tested wet or dry. Mix temperature influences the compressive strength of HSC by only 55% of the value obtained previously by others for NSC. The compressive strength of NSC & HSC rises significantly under short-term and long-term drying. However, HSC has a significantly smaller increase in compressive strength upon drying than NSC.
Specimens cured in water up to the age of 28 days followed by drying, exhibit sharply different effects of drying between compressive and tensile strengths. Short-term exposure to dry environment lowers the tensile strength of concrete. However, long-term drying with complete desiccation leads to an insignificant rise in tensile strength. This supports the conclusion, at this stage, that complete drying is not harmful to concrete tensile strength both for HSC and NSC.
Elevated temperature curing raises HSC compressive and tensile strengths. However, this influence becomes less significant for higher strength concrete.
Air curing beyond the age of 1 day has two contrasting effects on concrete strength. Compressive strength drop because of air curing becomes smaller for higher strength concrete. However, tensile strength drop due to air curing becomes more significant with stronger concrete. This contrast indicates the need for further research on the subject with a larger range of concrete strength.
Keywords: High/Normal-Strength Concrete, Elevated Temperature, Curing, Drying.

1 Introduction

High strength concrete (HSC) is now used increasingly in a variety of structures-tall buildings, highway bridges and pavement, prestressed concrete, severe environment structures, arch dams, piles for marine foundations, etc. (ACI Com. 363,1984). The inf-

Concrete in Hot Climates. Edited by M. J. Walker. © RILEM
Published by E & F N Spon, 2 - 6 Boundary Row, London SE1 8HN. ISBN 0 419 18090 7.

luence of environment on the properties of HSC is a current area of needed research (ACI Com. 363, 1987).

Temperature and/or humidity influence on the properties of normal strength concrete (NSC) has been studied for over 30 years. With its significantly different microstructure, HSC needs to be studied for the influence of environment on its behaviour and properties.

Microstructure studies indicate that with low temperature curing a relatively more uniform microstructure of the hydrated cement paste (especially the pore size distribution) accounts for the higher strength. Thus, mix temperature influence on concrete strength is important. For example, specimens cast, sealed and maintained at the indicated temperature for 2 hours, then stored at 21.1C curing (standard) until testing (wet), had significant differences in 28- , 90- and 180- day compressive strengths as follows in a descending order: 10C mix strength > 21.1C mix > 29.4C mix > 37.8C mix > 46.1C mix (Mehta 1986). In contrast, the modulus of rupture at 7 days for bond specimens cast and cured at 43.3C was twice that for prisms cured at 43.3C after casting at 21.1C (Alexander et al., 1965).

Recognition of the significant influence of mix temperature has led to concrete mix cooling in major projects. Recently for HSC, for example, liquid nitrogen was injected into the rotating drum mixing trucks. This successfully lowered mix temperature 18C (Ryell and Bickley 1987) or even as low as 8C (Springschmid and Breitenbucher 1987), during hot seasons.

Recently the authors (Samarai et al., 1987) concluded that mix temperature rise lowers 14-days compressive and tensile strengths of HSC (tested wet). In this paper the influence of mix temperature is studied for standard (wet) 28-day strengths.

Pihlajavaara (1972) reports that a 'thoroughly dry' concrete can have up to 50% higher compressive strength than a "thoroughly wet" concrete. Newman (1965) indicates that air drying can cause up to 30% increase in compressive strength. Since no tests are available for HSC, this project investigates the influence of mix temperature and drying on the compressive strength of HSC.

Walker and Bloem (1957) found that short drying periods led to significant drops in concrete flexural strength (modulus of rupture, f_r). For example, f_r dropped by 8% and 30% for drying periods of 30 min and 7 days respectively. This led them to conclude that the modulus of rupture is 'too sensitive to extraneous factors particularly moisture condition of specimens to permit its use as a basis for the acceptance or rejection of concrete in the field'.

On the other hand, one year drying at RH of 53-70% led to f_r having nearly the same value for wet and dry concrete (Walker and Bloem 1957). No research is available for the influence of drying on the more reliable split-cylinder tensile test, f_{ct}.

This work investigates the influence of mix temperature and drying on tensile splitting strength, f_{ct}, for HSC. Since the relative humidity (RH) for the Baghdad summer testing period is significantly less than its value indicated by Walker and Bloem (1957), shorter periods of drying have been used - mean Baghdad

66

summer RH is 14-37% (CIBS Tech. Memo., 1979).

Finally, tests are included to study the influence of elevated temperature curing and air curing on HSC compressive strength (f_{cu}) and f_{ct}. In order to separate these two parameters from mix temperature, these HSC specimens were all mixed at a normal temperature range (Table 1, groups 3 to 5).

2 Research significance

This paper reports on the influence of elevated mix temperature on the compressive and tensile strengths of HSC tested in

Table 1. Specimen group details

Group	Sub group	Specimen size & type	Mix temp range	Curing temp range	Type of curing	Test condition	Test age range (days)	Principal condition
1	A	150mm Cube	22.5C to 45C	23C to 29C	Water	Wet	28	Mix Temp
	B	150x300mm Cylinder						
2	A	150mm Cube	22.5C to 45C	23C to 29C	Water	Dry	56 to 81	Mix Temp & Post-Curing Drying
	B	150x300mm Cylinder						
3	A	75mm Cube	20C to 25C	50C **	Water	Wet	3 to 250	Elevated Curing Temp
	B	100x200mm Cylinder						
4	A	75mm Cube	20C to 25C	20C to 29C	Air	Dry	3 to 250	Lack Of Curing (Air Curing)
	B	100x200mm Cylinder						
5	A	75mm Cube	20C to 25C	20C to 29C	Water	Wet	3 to 250	Control Group For Groups 3 & 4
	B	150x300mm Cylinder						

*Cubes are for f_{cu} and cylinders are for f_{ct}

**All specimens from groups 3 whose test age exceeded 75 days were continuously cured at 23C to 29C beyond that age, until testing in wet condition.

standard (wet) condition at 28 days, as well as the combined
influence of mix temperature and drying up to the age of 81
days. In addition, the separate effects of 50C water curing &
air curing are also studied. This is the second paper in a programme intended to arrive at a better understanding of HSC properties in hot and dry environment.

3 Materials

3.1 Admixture
A locally available melamine formaldehyde condensate superplasticizer is used throughout the work. It complies with ASTM specification C494-82, Type F: water reducing high range admixture
(ASTM 1983). Specific gravity=1.1. Solid content=20% .

3.2 Cement
One brand of ASTM TYPE I cement is used throughout (ASTM 1983).

3.3 Sand
Sand of 2.4 fineness modulus has been used in this project.

3.4 Coarse Aggregate
Crushed gravel of 12 mm maximum size has been used.

4 Experemental Work

4.1 Mixing Concrete
Mixing sequence was:
1. coarse aggregate and sand added at the selected temperature
to the horizontal pan mixer; 2. one-minute mixing; 3. water
added at its selected temperature; 4. five-minute mixing; 5.
cement added; 6. superplasticizer added; 7. 2.5-minute mixing;
8. mix temperature measured.

  Mix temperature raising was by heating the water and/or the
aggregate, the latter being put in an oven for 24 hours prior to
mixing. Mix temperature lowering utilized refrigiration of the
ingredients and/or the addition of ice during mixing. Prior to
mixing the aggregate was re-weighed and any change in its water
content was compensated.

4.2 Pouring, Curing and Preparing Specimens
Compaction by vibrating table was in two layers with one minute
vibration per layer. Moulds were left in the laboratory overnight. Specimens were removed from the moulds within 24 hours of
pour time. Table 1 gives details of the curing regime for all 5
groups. Curing (water-or air-type) started in all cases on the
second day upon removal from the moulds.

4.3 Specimen Details and Testing
Each test result is a mean of 3 cubes or 3 cylinders for f_{cu} &
f_{ct} respectively. Table 1 details all 5 groups. Groups 1 & 2 are
from the same pour, with the principal variable being the mix
temperature, T. Group 1 gives f_{cu} & f_{ct} standard (wet) test res-

ults at 28-days-per ASTM (1983). Group 2 specimens are exposed to drying after the age of 28-days. Thus comparing groups 1 & 2 gives the influence of drying on f_{cu} & f_{ct}.

Groups 3 to 5 are from the same pour, group 5 being the control group. Thus, comparing groups 3 & 5 indicates the influence of 50C water curing up to the age of 75 days versus normal temperature curing, all specimens being tested in wet condition. Comparing groups 4 & 5 indicates the influence of air curing versus water curing-up to the maximum test age of 250 days.

5 Results and discussion

5.1 Effect of mix temperature on 28-day standard test compressive strength

The relationship between mix temperature and cube compressive strength, f_{cu}, for the specimens of subgroup/A tested wet at 28-days in accordance with ASTM specifications (1983) give the straight line relationship in Fig.1 following Eq.1.

$$\left.\begin{array}{l} f_{cu}\ (28) = 71.4 - 0.253\ T \\ \text{for the range: } 22.5\ C \le T \le 45\ C \end{array}\right\} \tag{1}$$

Eq.2 was previously found (3) for 14-day f_{cu}.

$$\left.\begin{array}{l} f_{cu}\ (14) = 69 - 0.187\ T \\ \text{for the range: } 13.8\ C \le T \le 44\ C \end{array}\right\} \tag{2}$$

Thus, for the limited number of tests made so far, the influence of mix temperature is about 35% greater for 28-days cube

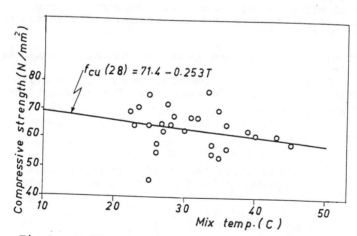

Fig.1.   Influence of mix temp. on wet specimen cube strength (f_{cu}).

compressive strength, f_{cu}, than 14-day results. Eq.1 gives a coefficient of variation (COV) of 10.7% for 28-day standard (wet) f_{cu} tests (Range of test results: 45.5-76.6 N/mm^2). The drop in compressive strength with rising mix temperature for HSC is in agreement with the trend indicated for NSC (ACI Com. 305, 1985 & Dodson and Rajagopalan 1979).

Dodson and Rajagopalan (1979) report a rate of 3.75 N/mm^2 drop per 10C rise, in mix temperature for 28-day cylinder compressive strength, f_c, for NSC. Eq.1 gives a drop of 2.53 N/mm^2 per 10C rise for 28-day cube compressive strength, f_{cu}, for HSC. Sarsam (1983) indicates a value of 0.82 for the ratio ($f_c^{'}/f_{cu}$). With this adjustment Eq.1 gives about 2.07 N/mm^2 drop in f_c per 10C rise in mix temperature for HSC. This is only 55% of the results reported for NSC (Dodson and Rajagopalan 1979). Thus for 28-day standard (wet) tests mix temperature has a significantly smaller influence on the compressive strength on HSC than NSC.

A probable contributor to this result is reported by Helland (1987) who concludes that the problem with high temperature rise in HSC is not proportional to the cement content. For example, the accumulated heat of reaction energies released after a period equivalent to 7 days at 20C (measured in KJ per kg of cement) for three mixes with w/c ratios of 0.30, 0.40 and 0.68 were in the ratios of 100, 119 and 137 respectively. This difference in heat of hydration per kg of cement becomes much more significant if it is coupled with the well known fact that the ratio of 7-day / long-term strength is higher for HSC than NSC (ACI Com. 363, 1984) e.g. the 37% increase (between w/c 0.3 and 0.68) in heat of hydration per kg of cement becomes even greater when converted to heat of hydration per unit compressive strength of concrete. Other factors may also contribute to the less harmful effect of mix temperature on HSC than NSC, such as the significant difference in microstructure between the two concretes (Mehta 1986).

5.2 Effect of mix temperature on 28-day standard test tensile strength

The relationship between mix temperature and split- cylinder tensile strength, f_{ct}, for the specimens of subgroup/B tested wet at 28-days in accordance with ASTM specifications (1983) gives the straight line relationship in Fig.2 following Eq.3 .

$$\left. \begin{array}{l} f_{ct} \; (28) = 6.74 - 0.034 \; T \\ \text{for the range: } 22.5 \; C \leq T \leq 45 \; C \end{array} \right] \tag{3}$$

Eq.3 gives a COV of 12.7% (Range of test results 4.67-7.87 N/mm^2). Eq.4 was previously found by Samarai et al. (1987) for f_{ct} at 14 days.

$$\left. \begin{array}{l} f_{ct} \; (14) = 6.68 - 0.273 \; T \\ \text{for the range: } 13.8 \; C \leq T \leq 44 \; C \end{array} \right] \tag{4}$$

Equation 3 indicates a 25% greater influence of mix temperature for f_{ct} at 28 days than at 14 days.

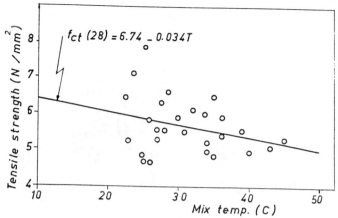

Fig.2.   Influence of mix temp. on wet specimen
         split cylinder strength (f_{ct}).

While 28-day modulus of rupture, f_r, drops significantly with mix temperature rise for NSC (Samarai et al., 1983 & Klieger 1985), no research is available for split-cylinder strength, f_{ct} (HSC or NSC). Thus, with the considerable difference in microstructure between HSC and NSC (Mehta 1986), future tests of mix temperature influence on tensile strength at different ages (7,-14,28,56,90 days, etc.) are indicated to compare the two types of concrete. Split-cylinder, f_{ct}, tests recommended since f_r tests are not reliable (Walker and Bloem 1957).

5.3 Effect of mix temperature on compressive strength of dry concrete
The relationship between mix temperature and f_{cu} for dry concrete specimens of subgroup 2A at the ages of 56 to 81 days gives the following straight line relationship.

$$f_{cu} \text{ (DRY)} = 80.3 - 0.283 \ \ T$$
$$\text{for the range: } 22.5 \ C \leq T \leq 45 \ C \tag{5}$$

Equation 5 gives a COV of 8.4%. (Range of test results: 54.1-82.7 N/mm^2). Comparing equations 5 and 1 leads to Eq.6 .
$$f_{cu} \text{ (DRY)} = 1.124 \ f_{cu} \ (28_{wet}) \tag{6}$$

Fig.3 shows no significant difference in results between the ages of 56 to 81 days representing drying after removal from water for 28 to 53 days at the low RH values in Baghdad. Average RH values are 19 & 47%, 13 & 34%, 12 & 32% and 13 & 33% respectively for the specimen drying months of May, June, July and August (CIBS Tech. Memo., 1979)-the first number is at 1500 Baghdad, mean time (BMT) and the second at 0600 BMT for each month. Neville (1981) indicates that desiccation was observed to

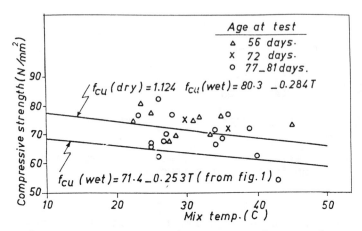

Fig.3.   Influence of mix temp. on dry specimen
         cube strength, f_{cu} (dry).

reach the depth of 75 mm for concrete in one month of drying—75 mm coincides with the maximum depth for subgroup 2A specimens.

Thus with the very low RH indicated above desiccation should have occured to all specimens. Pihlajavaara (1972) indicates that no hydration-strengthening occurs in concrete below RH of 80%. Thus no significant hydration- strengthening occured in the specimens after removal from water. Within the limited number of tests in this project, therefore, drying leads to 12.4% rise in f_{cu} for HSC. At the same time, the mix temperature influence is 12.4% greater for dry than wet HSC.

Table 2 compares the increase in compressive strength due to drying with other results (Pihlajavaara 1974 , Hsu and Slate 1963 & Popovics 1986). Popovics (1986) suggests a hypothesis that moisture gradient (outside layer drier than the inside) is the major contributor to higher strength, due to the shrinkage of the outside layer leading to lateral biaxial compression on the inside of the specimen. Popovics' hypothesis, based on 3-day drying does not fit the long-term drying results of Pihlajavaara (1974) or this work (Table 2), nor does it agree with indications by others (Pihlajavaara 1972 , Newman 1965 & Neville 1981) that drying itself (not the moisture gradient) raises the compressive strength of concrete.

Waters' (1955) hypothesis that in wet concrete water acts as a 'lubricant' that causes a strength drop is more appropriate for the influence of drying on strength. This lubricant hypothesis fits all results in Table 4 for compressive strengths with drying periods from 3 days to 3 years.

Table 2. Influence of drying on the compressive strength
of cementitious materials

Ref.	Material	Period of drying	Age of specimen	RH%	Change in compressive strength due to drying
This work	HSC	27-53 Days	56-81 Days	(+) 14-37%	+ 12.4%
Popovics (1986)	NSC	3 Days	28 Days	N.I.**	+ 17.8%
Pihlaja-vaara (1974)	Cement mortar (w/c=0.75)	3 Years	5 Years	7%	+ 50%
Pihlaja-vaara (1974)	Cement mortar (w/c=0.50)	3 Years	5 Years	7%	+ 30%
Hsu & Slate (1963)	Cement paste	3 Days	30-32 Days	53.70	+ 15%

+ Values are mean for the drying period, at 1500 and 0600
Baghdad mean time, respectively.
**N.I. = not indicated.

The 12.4% increase for HSC indicates that drying has a significantly smaller influence on HSC than NSC. This trend of smaller influence of drying for stronger material has already been
noted by Pihlajavaara (1974), which can be seen from Table 2
compare the 50% and 30% increases for mortars of w/c ratios of
0.75 and 0.50 respectively. The difference in microstructure of
HSC from NSC (Mehta 1986) is probably a major contributor to
this result. Another factor is likely to be the water content of
concrete. Since HSC starts with less water than NSC, HSC loses
less water upon drying, hence the smaller influence of drying on
the compressive strength of HSC.

5.4 Effect of mix temperature on tensile strength of dry concrete

The relationship between mix temperature and f_{ct}, for dry specimens of subgroup 2B at the ages of 56 to 81 days gives the straight line relationship in Fig.4 following Eq.7 .

$$f_{ct} (DRY) = 6.95 - 0.035 \ T \qquad (7)$$
$$\text{for the range: } 22.5 \ C \leq T \leq 45 \ C$$

Equation 7 gives a COV of 7.2%. (Range of test results:
5.24-6.53 N/mm^2). Comparing equations 7 and 3 leads to equation
8.

$$f_{ct} (DRY) = 1.031 \ f_{ct} (28_{wet}) \qquad (8)$$

73

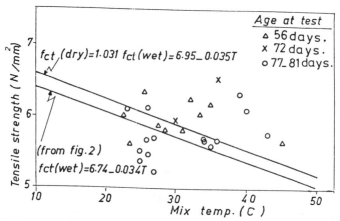

Fig.4.   Influence of mix temp. on dry specimen
          split cylinder strength, f_{ct} (dry).

Fig.4 shows no significant difference in results between the
ages of 56 to 81 days (drying of 28-53 days). As indicated
above, it may be assumed that the specimens are dry with little
or no hydration- strengthening after removal from water. The
3.1% increase due to drying is indignificant in itself. However,
it may be concluded that long-term drying is not detrimental to
HSC tensile strength, f_{ct}. Mix temperature influence on f_{ct} for
HSC is not significantly greater for dry than for wet concrete.-
While no research is available for the influence of drying on
split-cylinder strength, f_{ct}, the literature includes tests on
the modulus of rupture, f_{r}, and other tensile properties of
concrete, mortar and paste (Walker and Bloem 1957 ,
Pihlajavaara 1974 & Hsu and Slate 1963). Table 3 compares the
influence of drying on the tensile properties of cementitious
materials. Briquette tensile strength dropped by 58% and 19% for
cement paste and mortar respectively (Pihlajavaara 1974). In
3-day drying paste-aggregate bond strength dropped by 72% while
mortar-aggregate bond strength dropped by 77-93%. With third-
point loading, the modulus of rupture, f_{r}, for NSC had the following percentage differences from wet concrete: -30%, -30%, -6%
and +2% for drying periods of 7, 21, 84 and 357 days respectively at RH range of 53-70% (Walker and Bloem 1957), which is significantly higher than the RH values in Baghdad of 14-37% (CIBS
Tech. Memo. 4, 1979).
    Pihlajavaara (1974) found that 3-year drying (total age 5
years) at 7% RH led to 70% and 60% increase in f_{r} (centre- point
loading of mortar beams) with w/c ratio of 0.75 and 0.50 respectively. As found previously for compressive strength, this indicates that the stronger material is less sensitive to drying.
    From the results discussed above, previous research (Walker
and Bloem 1957 & Hsu and Slate 1963) clearly indicates that

Table 3. Influence of drying on the tensile properties of cementitious materials

Ref.	Material	Period of drying	Age of specimen	RH%	Type of strength	Change in strength
This work	HSC	27-53 Days	56-81 Days	14-37% (+)	f_{ct}	+ 3.1%
Walker et al. (1957)	NSC	7 Days	14 Days	N.I.**	f_r	+ 30%
Walker et al. (1957)	NSC	21 Days	28 Days	N.I.	f_r	- 30%
Walker et al. (1957)	NSC	84 Days	91 Days	N.I.	f_r	- 6%
Walker et al. (1957)	NSC	357 Days	364 Days	N.I.	f_r	+ 2%
Pihlaja-vaara (1974)	Mortar (w/c=.75)	3 Years	5 Years	7%	f_r	+ 70%
Pihlaja-vaara (1974)	Mortar (w/c=.50)	3 Years	5 Years	7%	f_r	+ 60%
Hsu & Slate (1963)	Mortar	3 Days	30-32 Days	53.70	Tensile briquitte	- 19%
Hsu & Slate (1963)	Cement paste	3 Days	30-32 Days	53.70	Tensile briquette	- 58%
Hsu & Slate (1963)	Mortar	3 Days	30-32 Days	53.70	Mortar-agg.bond	- 77% to - 93%
Hsu & Slate (1963)	Cement paste	3 Days	30-32 Days	53.70	Paste-agg.bond	- 72%

+ & ** - See footnote to Table 2.

short-term drying is detrimental to the tensile properties of cement paste, mortar and NSC as determined by paste-aggregate bond, mortar-aggregate bond, briquette tension and modulus of rupture tests. Only extremely low RH (7% of Pihlajavaara 1974) for a very long drying period (3 years) led to a significant increase in the modulus of rupture of mortar. However, for NSC (Walker and Bloem 1957) 1-year drying led to an insignificant change in the modulus of rupture, f_r (+2%) just as is found in this work that for HSC 28-53 drying days led to an insignificant change in split-cylinder tensile strength, f_{ct} (+3.1%).

Thus, based on this work and Walker and Bloem's (1957) results, it may be concluded that long-term drying is not harmful to the tensile strength of NSC or HSC. With the available experimental evidence so far, the significant increases reported by Pihlajavaara (1974) for mortar cannot be used to conclude that concrete benefits significantly in tensile strength from long-term drying. However, future companion tests on HSC and NSC under similar drying conditions should shed more light on this matter.

5.5 Effect of elevated temperature curing on strength
Fig.5 compares subgroups 3A & 5A for the influence of 50C curing on compressive strength [$f_{cu}(50)$] versus normal temperature curing [$f_{cu}(NORM)$]. Both subgroups have been tested in a wet condition. The following straight line relationship is obtained.

$$f_{cu}(50) = 0.908 \ f_{cu}(NORM) + 9.1 \qquad (9)$$

Eq.9 gives a COV of 11.1%. The range of test results is 24.5 to 63.9 N/mm² for $f_{cu}(50)$ and 20.6 to 64.0 N/mm² for $f_{cu}(NORM)$.

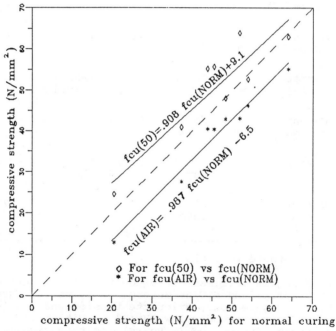

Fig.5. Influence of air curing and
      elevated temperature curing on
      the cube strength,fcu.

Eq.9 indicates that the enhancing influence of 50C curing becomes less significant as concrete strength rises.

Fig.6 compares subgroups 3B & 5B for the influence of 50C curing on tensile strength [$f_{ct}(50)$] versus normal temperature curing [$f_{ct}(NORM)$]-both subgroups being tested in a wet condition. The following straight line relationship is obtained.

$$f_{ct} (50) = 0.848 f_{ct} (NORM) + 1.17 \qquad (10)$$

Eq.10 has a COV of 12.6%. The range of test results is 3.0 to 7.3 N/mm^2 for $f_{ct}(50)$ and 2.6 to 7.6 N/mm^2 for $f_{ct}(NORM)$. As with compressive strength, tensile strength enhancement from 50C curing becomes less significant as concrete strength rises.

ACI Com. 363 (1984) indicates that higher strength concrete gains a greater proportion of its strength at earlier ages, when compared with lower strength concrete. 50C curing for $f_{cu}(50)$ & $f_{ct}(50)$ is started after a day in the moulds the same as normal temperature curing for $f_{cu}(NORM)$ & $f_{ct}(NORM)$. Thus in these two groups (3 & 5) curing beyond the age of 1 day would have a smaller proportion of concrete strength to influence. This may explain the fact that Eq.9 & 10 show a dropping influence of elevated temperature curing when concrete strength rises.

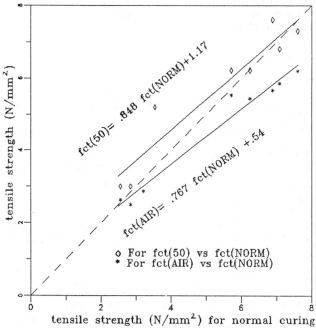

Fig.6. Influence of air curing and
elevated temperature curing on
the split cylinder strength,fct.

5.6 Effect of air curing on strength

Fig.5 compares subgroups 4A & 5A for the influence of air curing on compressive strength [f_{cu}(AIR)] versus normal water curing [f_{cu}(NORM)]. The following straight line relationship is obtained.

$$f_{cu} \text{ (AIR)} = 0.987 \, f_{cu} \text{ (NORM)} - 6.5 \tag{11}$$

Eq.11 give a COV of 6.53%. The range of test results is 13.0 to 55.2 N/mm^2 for f_{cu}(AIR). These results indicate that air curing causes a drop in compressive strength, when compared to normal wet curing. This influence, however, drops as a proportion of strength with rising concrete compressive strength.

Fig.6 compares subgroups 4B & 5B for the influence of air curing on tensile strength [f_{ct}(AIR)] versus normal water curing [f_{ct}(NORM)]. The following straight line relationship is obtained.

$$f_{ct} \text{ (AIR)} = 0.767 \, f_{ct} \text{ (NORM)} + 0.54 \tag{12}$$

Eq.12 has a COV of 6.47%. The range of test results is 2.25 to 6.20 N/mm^2 for f_{ct}(AIR). In contrast with Eq.11, Eq.12 indicates that higher strength concrete is harmed by air curing to a larger extent than lower strength concrete. Thus, further research is indicated in this respect at a larger range of tensile strength, f_{ct}.

6 Conclusions

Based on results obtained from tests in this work and by other researchers, the influence of environment on the properties of concrete (temperature and/or drying of concrete) is as follows.

(a) Mix temperature influence on 28-day standard (wet) compressive strength of HSC is only 35% greater than on 14-day tests.
(b) The influence of mix temperature on 28-day wet specimen compressive strength is significantly smaller for HSC than NSC-ratio of HSC/NSC influence is 55%.
(c) Mix temperature influence on 28-day standard (wet) split-cylinder strength, f_{ct}, of HSC is only 25% greater than on 14-day tests.
(d) In concrete cured in water to the age of 28 days, subsequent drying has a significantly smaller influence on HSC than NSC. The difference in microstructure between the two types of concrete probably contributes to this result.
(e) In concrete cured in water up to 28-day age, subsequent rise in compressive strength due to drying may be explained by a hypothesis originally proposed by Waters (23). This suggests that, in wet concrete, water acts as a lubricant which causes the drop in f_{cu}. Thus with its smaller water content

HSC undergoes a significantly smaller strength change under dry environment than NSC.
(f) Mix temperature has 12.4% greater influence on the compressive strength of dry than wet HSC.
(g) Short-term exposure to dry environment of wet concrete, which leads to incomplete drying, is harmful to its tensile strength.
(h) Long term complete drying is not harmful to the tensile strength of both HSC and NSC. In contrast with compressive strength, however, no conclusive evidence is available so far that long term drying significantly raises the tensile strength of concrete.
(i) Elevated temperature (50 C) curing raises the concrete compressive and tensile strengths (f_{cu} & f_{ct}). However, this effect becomes less significant as concrete strength rises.
(j) Air curing beyond the age of 1 day causes a drop in concrete compressive strength. This drop becomes less pronounced with higher strength concrete.
(k) In contrast with compressive strength, however, tensile strength (f_{ct}) drop caused by air curing (beyond the age of 1 day) is more significant for higher strength concrete. This results indicates the need for further investigation of the influence of air curing on f_{ct} with a larger range of concrete strength.

7 Acknowledgement

The authors wish to thank the National Centre for Construction Laboratories for providing the funds and facilities for the research. The Building and Construction Engineering Department, University of Technology is thanked for providing time and encouragement for one of the authors. Thanks are also due to the College of Engineering, University of Baghdad, for the use of their laboratory facilities.

8 References

ACI Committee 363 (1984) State-of-the-art report on high strength concrete. **ACI Journal.**, Vol.81, No.4, pp.364-411.
ACI Committee 305 (1985) Hot weather concreting, ACI 305R- 77, **Manual of Concrete Practice**, Part 2, Amer. Conc. Inst.
ACI Committee 363 (1987) Research needs for high strength concrete. **ACI Materials Journal.**, Vol.84, No.6, pp.559-661.
Alexander, K.M., Wardlaw, J., and Gilbert, D. (1965) Aggregate-cement bond, cement paste strength and the strength of concrete, Proc.of an International Conf. on **The Structure of Concrete**, Imperial College, London, pp.59-81.
Annual Book of ASTM Standards (1983), Vol.04.02, Concrete and **Mineral Aggregates**.
CIBS Technical Memoranda 4 (1979) **CIBS Design Notes for the Middle East**, London, 61 pp.
Dodson, C.J., and Rajagopalan, K.S. (1979) Field tests verify temperature effects on concrete strength, Concrete

International: Design and Construction, Vol.1, No.12, pp.26-30.

Helland, S. (1987) Temperature and strength development in concrete with w/c less than 0.40, Proc. of Symp. on **Utilization of High Strength Concrete**, Tapir Publishers, Stavanger, Norway, pp.473-485.

Hsu, T.T.C., and Slate, F.O. (1963) Tensile bond strength of aggregate and cement paste mortar, **ACI Journal**, Vol.60, pp.465-485.

Klieger, P. (1985) Effect of mixing and curing temperature on concrete strength, **ACI Journal**, Vol.54, pp.1063-1081.

Mehta, P.K. (1986) **Concrete: Structure, Properties and Materials**, Prentice-Hall Inc., 450 pp.

Neville, A.M. (1981) **Properties of Concrete**, Pitman Publishers, 779 pp.

Newman, K. (1965) The structure and properties of concrete- an introductory review, Proc. of an International Conf. on **The Structure of Concrete**, Imperial College, London, pp.xiii-xxxii.

Pihlajavaara, S.E. (1972) An analysis of the factors exerting effect on strength and other properties of concrete at elevated temperatures, Amer. Conc. Inst. Publication No.SP-34, **Concrete for nuclear reactors**, V.1, pp.347-354.

Pihlajavaara, S.E. (1974) A review of some of the main results of a research on the ageing phenomena of concrete: effect of moisture conditions on strength, shrinkage and creep of mature concrete, **Cement and Concrete Research**, Vol.4, pp.761-777.

Popovics, S. (1986) Effect of curing method and final moisture condition on compressive strength of concrete, **ACI Journal**, V.83, No.4, pp.650-657.

Ryell, J., and Bickley, J.A. (1987) Scotia Plaza: high strength concrete for tall buildings, Proc. of Symp. on **Utilization of High Strength Concrete**, Tapir Publishers, Stavanger, Norway, pp.641-653.

Samarai, M., Popovics, S., and Malhotra, V.M. (1983) Effects of high temperature on the properties of hardened concrete, **Transportation Research Record 924**, TRB, National Academy of Sciences, Washington, D.C.

Samarai, M.A., Sarsam, K.F., and Al-Khafagi, M. (1987) Influence of mixing temperature on the properties of high strength concrete, Proc. of Symp. on **Utilization of High Strength Concrete**, Tapir Publishers, Stavanger, Norway, pp. 443-471.

Sarsam, K.F. (1983) **Strength and deformation of structural concrete joints**, Ph.D. Thesis, University of Manchester Institute of Science and Technology, 340 pp.

Springenschmid, R., and Breitenbucher, R. (1987) Technological aspects for high-strength-concrete in thick structural members, Proc. of Symp. on **Utilization of High Strength Concrete**, Tapir Publishers, Stavanger, Norway, pp.487-496.

Walker, S., and Bloem, D.L. (1957) Studies of flexural strength of concrete, Part 3; effects of variations in testing procedures, ASTM, Proc., Vol.57, pp.1122-1142.

Waters, T. (1955) The effect of allowing concrete to dry before it has fully cured, **Magazine of Concrete Research**, Vol.7, No.20, pp.79-82.

8 EFFECT OF EVAPORATION OF WATER FROM FRESH CONCRETE IN HOT CLIMATES ON THE PROPERTIES OF CONCRETE

Z. BERHANE
Department of Civil Engineering, Addis Ababa University,
Ethiopia

Abstract
Evaporation of water from fresh concrete has both positive and negative effects on the properties of concrete. Excessive evaporation will adversely influence the workability of fresh concrete thereby encouraging re-tempering, which in turn, will increase the water-cement ratio and as a result the quality of the hardened concrete will be lowered. In hot-dry climatic condition, where the relative humidity is very low, plastic shrinkage cracking is likely to occur. On the other hand, loss of water from fresh concrete, while the fresh concrete is still plastic, could help densify (compact) the fresh concrete because of collapsing of the water channels thereby bringing down the void content of the concrete and also lowers the water-cement ratio thereby improving the properties of the hardened concrete. The paper discusses both positive and negative effects of evaporation of water from fresh concrete and points out precautions to be taken in order to attain optimum positive influences of evaporation of water from fresh concrete.
Keywords: Hot-humid, hot-dry, Bleeding, Evaporation, Densification, Plastic shrinkage cracking, Strength, Drying shrinkage, Creep.

1 Introduction

At present evaporation of bleeding water from fresh concrete in hot climates is mainly known for its adverse effects (slump loss, plastic shrinkage cracking, deficiency in strength at later ages, etc.). However, its positive

Concrete in Hot Climates. Edited by M. J. Walker. © RILEM
Published by E & F N Spon, 2 - 6 Boundary Row, London SE1 8HN. ISBN 0 419 18090 7.

effects such as densification of fresh concrete(1), which lowers the water-cement ratio thereby increasing the strength of concrete, and bringing down drying shrinkage and creep, improving the watertightness and upgrading its durability(2) are not usually taken into account. Nevertheless, one should keep in mind that such advantages could only be achieved when evaporation of water does not attain values which will cause the harmful effects mentioned earlier(2).

2 Behaviour of fresh concrete

As it is well known, as soon as cement particles come in contact with water a sharp and immediate chemical reactions take place. However, these initial reactions do not last long. Then a fast deceleration in chemical reaction takes place leading to the so called dormant period. The dormant period lasts 40 to 120 minutes at normal temperature(2). Experimental results show that the duration of the dormant period is reported to be shortened by elevated temperature such as encountered in hot climates(3,4). Obviously, the shortening of the dormant period in hot climates will have its own practical implication such as loss of slump and workability.
   During the dormant period, fresh concrete could be assumed to be composed of inert different size particles. At this stage only interparticle attractive and repulsive forces are at work. Obviously, during the dormant period interparticle gaps exist, and settlement and bleeding take place. Water being the lightest component of the fresh concrete mixture tends to accumulate on the surface, while the heavier ingredients tend to settle at the bottom. It is interesting to note that water flows around the particles individually, but not around clusters of particles(2). It is also reported that at normal condition the initial rate of bleeding of fresh concrete is almost constant lasting 15 to 30 minutes, and thereafter the rate diminishes, reach zero within an hour and a half(2). However, eventually the interparticle gaps become bridged with Portland cement hydration products.

3 Evaporation of water from fresh concrete

Evaporation is the conversion of water from the liquid or solid state into gaseous at a temperature below boiling point and its diffusion into the atmosphere(5). It occurs wherever a water surface is exposed to overlying air which remains unsaturated(6). In case of fresh concrete the water surface is created due to bleeding water being accumulated on the surface of the fresh concrete.

As it is well known, an open water surface is an open water interface, across which there is an exchange of water molecules in both directions. Even when the water body is progressively evaporating into the adjacent air, it is the net movement of water molecules into the air one measures as evaporation as some water molecules will still be transferring from the air to the water body(6). Wind (air movement) serves to distribute water vapour vertically and horizontally which results in drier air to be brought down into contact with the water surface and saturated air at or near the water surface to be lifted up into the atmosphere.

It seems that at present there is no an exact method for measuring the rate of evaporation of water from a water surface. The presently used different methods are reported to yield only estimates(5,6,7).

Empirical formulae for estimating rate of evaporation are derived using meteorological data. There are a number of methods used to estimate evaporation from a body of water or wet surfaces. The main ones are: i) the water-budget method, ii) the heat balance or energy budget method and iii) the aerodynamic approach.

As it is well known, the sun at a temperature of about 6,000 K° (5727°C) is the source nearly all our energy. The earth intercepts an infinitesimally small part of the sun's out put 5x10^{-10}%(8). Only a portion of the sun's radiation reaches the earth's surface as direct radiation; the remainder being reflected, absorbed, or scattered by the atmosphere(9).

The water-budget method is mainly used to measure evaporation from a reservoir and is said to be simple(10). After allowing for any change in the volume of water stored in the reservoir, evaporation is computed as the difference between inflow and outflow. The formula has the following form:

$$E = I + P - O - S \qquad (1)$$

in which I is the volume of inflow to the reservoir, P is the amount of precipitation falling on the surface of the reservoir, O is the volume of outflow from the reservoir, S is change in the volume of water contained in the reservoir and E is the volume of evaporation from the reservoir. The application of the formula to measure the evaporation of water from fresh concrete seems remote.

The energy budget approach assumes that if radiation available for evaporation process is known, the evaporation rate may be estimated using the following formula:

$$E = R_n / L(1 + \beta) \qquad (2)$$

where E is the rate of evaporation, Rn is the net radiation flux, L is the latent heat of evaporation and

β is the ratio between flux to the atmosphere and the latent heat.

The aerodynamic approach was originally developed by John Dalton(11) and is usually and wrongly quoted as the Menzel's formula in concrete technology. The formula has the following form:

$$E = (e_w - e_a) \ f(u) \tag{3}$$

where e_w is the vapour pressure at the evaporating surface, e_a is the vapour pressure in the atmosphere above the evaporating water surface and $f(u)$ is a function of horizontal wind velocity. Since its original formulation of equation (3) has been modified to suit certain conditions and the one given below is the formula adapted to be used with a wind speed measurement made at two meters above the water surface.

$$E = 0.4(e_w - e_a) \ (1 + 0.17V) \ mm/day \tag{4}$$

where V is the velocity of wind.

The formula which is commonly used in concrete technology is the one given below.

$$E = 0.44(e_w - e_a) \ (0.253 + 0.09V) \tag{5}$$

where E is the rate of evaporation in lb./ft^2 per hour (1 lb./ft^2/hr. = 4.88 kg/m^2/hr.) and e_w and e_a are in psi and V in miles/hr.

However, the concrete technologist is not only interested in the evaporation rate and amount from fresh concrete surface covered with bleeding water, but also in the rate of evaporation of water from the interior of fresh concrete without bleeding water on its surface. In fact, evaporation of water without bleeding water on the fresh concrete is assumed to be the main cause of plastic shrinkage cracks(12).

Some have come to conclusions that Menzel's formula may not be applicable for estimating evaporation rate at any stage of fresh concrete at any climatic conditions(10,15).

If we assume an analogy of fresh concrete and wet soil(10) three stages will take place during the process of evaporation. In the first stage evaporation occurs from bleeding water surface. Under such conditions evaporation is controlled by the factors which control free water surface evaporation. After the bleeding water has been consumed because of evaporation, the second stage begins when the fresh concrete surface is kept moist by capillary ascent of water from beneath. Under such conditions it is easier for the capillary water supply to keep pace with the water lost by evaporation when the evaporative power of air is small and vice versa. In the latter case the concrete surface gets drier and plastic shrinkage cracks may occur. The third stage begins with the formation of a dry concrete surface of a certain depth. In this stage the water loss is assumed to be controlled completely by the temperature gradient within the concrete.

It is reported that evaporation from wet soil is influenced by many factors such as the character of the soil itself, texture, grain size, grain size distribution, surface roughness and colour(6,9). However, albedo of the soil surface is reported to be a major factor. As it well known, albedo of a surface is the degree of reflecting incoming solar radiation which determines the amount of energy which may remain in a material for the evaporation process. To the knowledge of the writer there are no published data on albedo of fresh concrete surfaces at present. All above mentioned factors might hold true for fresh concrete.

Thornthwaite and Holzman(9) state that when the surface has become dry or partially dry, less evaporation occurs from the soil than from a water pan. Even though, the subsoil is moist, capillary action cannot supply the surface with water at a rate comparable to the evaporation from the surface of a body of water. Laboratory measurements of rate of evaporation of water at different climatic conditions show the same trend is true with fresh mortar and concrete(10).

The presently well known main factors, which are assume to have a pronounced influence on the rate of evaporation of water from fresh concrete surface, are: i) the ambient temperature, ii) the initial temperature of the fresh concrete, iii) the magnitude of the prevailing relative humidity and iv) the wind velocity(10,12,13). A combination of elevated temperature, low relative humidity and strong wind will cause a very high rate of evaporation. It is interesting to note that when fresh concrete is exposed immediately after mixing to hot-dry (desert) climate the very high rate of evaporation will cause a sharp drop in the temperature of the fresh concrete whereas in a hot-humid (tropical) climate, because of the negligible cooling effect of the low rate of evaporation the temperature of the fresh concrete will increase significantly above the initial one(14). Obviously, such an elevated temperature will adversely affect the micro structure of the cement hydration products, which in turn, will adversely influence the properties of concrete.

4 Effect of evaporation on the properties of hardened
  concrete

As it was stated earlier, if evaporation takes place while the concrete is still plastic and the walls of the water

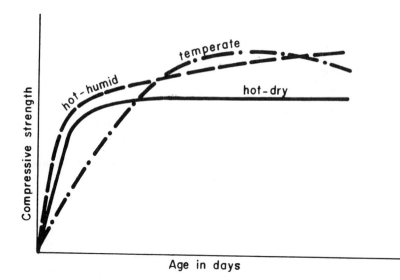

Fig. 1. Schematic representation of the influence
of different climates on the strength
development of concretes.

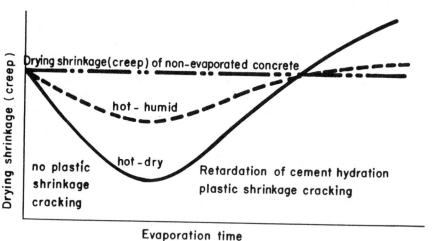

Fig. 2. Schematic diagram showing the effect of
evaporation of water from fresh concrete
surface on the drying shrinkage (creep)
of concrete in hot climates.

channel can collapse and close the channels, the fresh concrete is densified and at the same time a reduction in the water-cement ratio of the concrete is attained, which in turn, will have positive effects on the strength, drying shrinkage and creep of the hardened concrete (see Figs. 1 and 2). Above a certain limit, a high rate of evaporation will decrease workability thereby encouraging retempering and might cause plastic shrinkage cracks(2), which in turn, will have adverse effects on the mechanical properties and durability of the concrete and might retard and even arrest cement hydration.

Even though, concrete cast at and exposed to hot-humid climate usually shows deficiencies in strength at about 28 days, it seems that concrete exposed for its life to such a climate might have equal to or higher strength than that prepared and exposed for its life to normal climatic conditions(16) (see Fig. 1). Recent investigation results have confirmed this assumption(17).

5 Concluding remarks

Though evaporation of water from fresh concrete is presently well known mainly for its adverse effects on both fresh and hardened concrete, it has also positive influences on the properties of concrete. However, before one takes advantage of such positive effects further studies have to be done by exposing fresh concrete to different hot climates for different length of time in order to pin point optimum exposure times which will give possible best results.

Acknowledgments

The author wishes to thank Taywood Engineering Ltd (TEL), the Centre for Materials and the Constructed Environment (MACE) at the Imperial College, London and the UK Science and Engineering Research Council (SERC) for making it possible for him to visit Imperial College (where a portion of the present work was done) as a Senior Visiting Fellow while on sabbatical leave from Addis Ababa University.

6 References

1. Powers, T. C. (1968) "Properties of fresh concrete," John Wiley & Sons, New York, pp. 598-599.
2. Zawde Berhane (1992) "Behaviour of concrete in hot climates," Materials and Structures, Vol. 25, No. 147, pp. 157-162.
3. Danielsson, U. (1966) "Conduction calorimeter studies

of the heat of hydration of a Portland cement," Swedish Cement and Concrete Research Institute, Stockholm, Sweden, pp. 1-121.

4. Manfore, G. E. and Ost, B. O. (1966) "An isothermal conduction calorimeter for study of the early hydration reactions of Portland cements," PCA J. Res. and Dev. Lab., Vol. 8, No. 2, pp. 13-20.

5. Wiesner, C. J. (1970) "Hydrometeorology", Chapman & Hall Ltd, London, pp. 71.

6. Sumner, G. (1988) "Precipitation: process and analysis," John Wiley and Sons, New York, pp. 234-34.

7. Lancelet, R. K., Kohler, M. A. and Paulhus (1988) "Hydrology for engineers," McGraw Hill Book Company, New York, pp. 126-163.

8. Strahler, A. N. (1973) "Introduction to physical geography," John Wiley and Sons, New York, pp. 10.

9. Thornthwaite, C. W. and Holzman, B. 1939) "The determination of evaporation from land and water surfaces", Monthly Weather Report, U. S. Department of Agriculture, Weather Bureau, Vol. 69, pp. 1-10.

10. Zawde Berhane (1984) "Evaporation of water from fresh mortar and concrete at different environmental conditions," ACI J. Vol. 81, No. 6, pp. 560-565.

11. Dalton, J. (1802) "Experimental essay on the constitution of mixed gases; on the force of steam or vapour from waters and other liquids, both in a torrcellian vacuum and in air; on evaporation; and on the expansion of gases by heat", Proc. Manch. Lit. Phil. Soc., Vol. 5, pp. 535-602.

12. Lerch, W. (1957) "Plastic shrinkage," ACI J., Vol.53, pp. 797-802.

13. Ravina, D. and Shalon, R. (1968) "Plastic shrinkage cracking," ACI J., Vol. 65, pp. 282-292.

14. Author's closure of discussion of Ref. No. 10 (1985) ACI J., Vol. 82, pp. 931-932.

15. Hasanain, G. S., Khallaf, T. A. and Mahmood, K. (1989) "Water evaporation from freshly placed concrete surfaces in weather", Cem. Conc. Res., Vol. 19, pp.465-475.

16. Zawde Berhane (1983) "Compressive strength of mortar in hot-humid environmental condition", Cem. Conc. Res., vol. 13, pp. 225-232.

17. Mustaf, M. A. and Yusof, K. M. (1991) "Mechanical properties of hardened concrete in hot-humid climate", Cem. Conc. Res.. Vol. 21, pp. 601-613.

9 STRENGTH DEVELOPMENT AND MICROSTRUCTURE OF CEMENT AND CONCRETE PRECOOLED WITH LIQUID NITROGEN

S. NAKANE, H. SAITO and T. OHIKE
Technical Research Institute, Obayashi Corporation,
Tokyo, Japan

Abstract
Injecting liquid nitrogen(LN$_2$) into fresh concrete in a truck agitator is one method of precooling concrete. However, since the temperature of LN$_2$ is $-196°C$, the influence on hydration of cement
subjected to this cryogenic temperature will be a matter of concern
at times.
   In this study, three precooling measures were taken such as cooling by LN$_2$, air–cooling materials before mixing, and adding ice chips
in place of a part of mixing water, and the differences in strengths,
microstructures and rate of hydration of hardened cement paste in
each case were investigated. The results were as follows: (1) The
difference in cooling methods when cooling concrete to the same degree
dose not influence strength, microstructure, and rate of hydration of
cement paste. (2) Even when cooled down to almost $0°C$ by LN$_2$, both
concrete and cement paste show good strength development properties.
Keywords: High Strength Mass Concrete, Liquid Nitrogen, Pre–cooling,
Strength Development, Microstructure.

1 Introduction

Previous experiments have cleary shown that by precooling mass
concrete placed in summertime, it is not only possible to control
thermal cracking, but also improve the strength development of
concrete in structures.[1][2][3] Cooling media used for pre–cooling
include cold water, ice, and, most recently, liquid nitrogen.
   This report deals with a method of cooling concrete in a ready-
mixed concrete agitator truck by injecting directly liquid nitrogen.
The temperature of the liquid nitrogen is about $-196°C$ and, although
just for a moment, it lowers the temperature of the cement particles
in the concrete to an extremely low level. Some members of the
industry are concerned that this process might adversely affect the

Concrete in Hot Climates. Edited by M. J. Walker. © RILEM
Published by E & F N Spon, 2 - 6 Boundary Row, London SE1 8HN. ISBN 0 419 18090 7.

later hydration process.This report describes, therfore, the result of experimental studies conducted to determine if using liquid nitrogen as a cooling medium affects the hydration of cement paste. The following two experiments were performed.

    1) Macroscopic verification of the effect of precooling mass concrete with liquid nitrogen. (concrete experiment)
    2) An investigation of the effect of differences in the cooling method and degree of cooling on the strength development and the microstructure of cement. (cement paste experiment)

2 Verification of the effect of precooling mass concrete with liquid nitrogen

This experiment was conducted to verify experimentally that it is possible to improve the strength of high–strength mass concrete by precooling it with liquid nitrogen.

2.1 Outline of the Experiment

The concrete used in the experiment was high–strength mass concrete with specified design strength 420 – 450 kgf/cm^2. Moderate–heat cement containing 20% fly ash was used. Table 1 shows the mix proportion of concrete.

    The test conditions that made up the overall experiment are shown in Table 2.

Table 1. Mix proportion of concrete

W/C (%)	S/A (%)	Cement (kg/m^3)	Water	Fine aggregate	Coarse aggregate	Admixture
40	42	430	172	710	1004	4.30

Table 2. Experiment conditions (concrete)

Experiment cord	Precooling condition Method	Range	Casting temperature	Curing conditions	Temperature hysteresis
A	----	----	30℃	Standard	I
B	----	----	30℃	High tem-	II
C	LN$_2$	30→15℃	15℃	perature	III
D	LN$_2$	30→ 0℃	0℃	hysteresis sealing	III

Tests A and B were performed to investigate the control cylinder strength (standard water curing) and the structural strength under temperature hysteresis without precooling, respectively.

Test C estimated the temperature hysteresis in structures when the concrete was precooled with liquid nitrogen. Test D was similar to test C, but this was used to estimate the conditions when the concrete was partially frozen as a result of accidental uneven distribution of the liquid nitrogen in the ready–mixed concrete agitator truck during precooling.

Figure 1 shows the curing temperature control chart in each case. Temperature hysteresis curing refers to curing condition reproduced in a tunk under which the structural element is supposed to undergo when concrete is cast in the summer.

2.2 Results and discussion
Figure 2 illustrates the strength development that accompanied aging.

The results of A and B indicate that the standard water curing strength increases as the concrete ages, but as already pointed out, the temperature hysteresis curing strength dose not appear to increase in the long term, despite the fact that considerable strength development at the initial stage.

It is clear that when precooling was performed the temperature hysteresis curing strength was improved. In other words, the strength in case C, which had a low placement temperature, was approximately 60kgf/cm^2 greater than B after 13 weeks of aging.

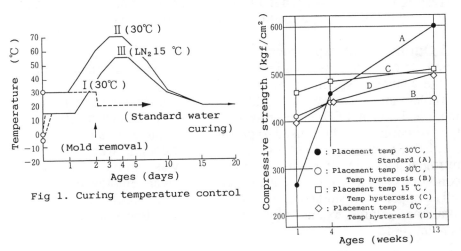

Fig 1. Curing temperature control

Fig 2. Degree of cooling, curing conditions and strength development

91

The above results can be said to have clearly reproduced the characteristics of the strength development of high–strength mass concrete placed in summertime. [1] [2] [3]

In case D, based on the premise that partial excessive cooling occurred in a ready–mixed concrete truck, the strength was as well as case B concrete by the 4th week. Furthermore it was as well as case C concrete strength in the 13th week, demonstrating that the effects of the freezing are not revealed macroscopically.

This enables to conclude that these experiments have verified that, as has already been asserted, precooling with liquid nitrogen effectively improves the structural strength development of mass concrete.

3 An investigation of the effect of the cooling methods and degrees of cooling on the strength and the microstructure of cement paste

Through the experiments described in section 2 above, it was possible to verify the effectiveness of precooling mass concrete with liquid nitrogen. There are, however, some who are concerned that the temperature of the cement particles will be lowered to extremely low levels momentarily or temporarily by the liquid nitrogen.

So this experiment examines the strength development and microstructure of cement paste under various cooling methods and degrees of cooling to determine the effect of these changes.

Testing material was changed from concrete to cement paste so that it would be easier to analyze the microstructure.

3.1 Outline of the experiment
The type of cement used and water/cement ratio are identical to those of the concrete experiment in section 2.

Table 3 indicates the test conditions that composed the experiment. PB, PC and PD represent a study of differences caused by the type of cooling medium used to perform the same degree of cooling. PA, PD, PE and PF represent of the degree of cooling with liquid nitrogen, including freezing. After the mold was removed, the same conditions as those for standard water curing were employed.

The experiment studied compressive strength, pore distribution, total pore volume, extent of hydration, and the microstructure of the cement paste.

Table 3. Experimental conditions (cement paste)

Experiment cord	Precooling condition		Casting temperature	Curing conditions
	Method	Range		
PA	----	----	30°C	
PB	Air	30→15°C	15°C	Standard
PC	Ice	30→15°C	15°C	water
PD	LN$_2$	30→15°C	15°C	curing
PE	LN$_2$	30→ 0°C	0°C	
PF	LN$_2$	30→Freezing	0°C	

3.2 Results and discussion

3.2.1 Strength development

Figure 3 compares strength development for various cooling media.

The figure illustrates that there is little difference between the strength development of cement paste subjected to three different cooling methods: air–cooling the raw materials before mixing, cooling the materials with ice as they are mixed, or cooling the mixed cement paste with liquid nitrogen.

This demonstrates that using liquid nitrogen as the cooling medium has no effect on the hydration of the cement.

Figure 4 shows a comparison of the extent of cooling (includng over–cooling) using liquid nitrogen with strength development patterns.

The figure reveals that the initial strength development of both PD which was cooled to 15°C and PE which was cooled to 0°C, were smaller than that of PA which was not cooled. After 13 weeks of aging, however, the strengths of PD and PE shows highest strength.

The strength development of over–cooled samples (PF) that were frozen for an hour or more was a little lower than that of PD and PE, both of which were subjected to normal cooling. Compared with the sample that was not cooled (PA), however, the strength development of PF was a little slow up until aging for 28 days, but at 13 weeks, it had achieved almost equal strength.

These results show that cooling without freezing has absolutely no adverse effect upon strength, even when the material was cooled to 0°C with liquid nitrogen. Furthermore the strength development of the cement paste that was temporarily over–cooled was not effected very much.

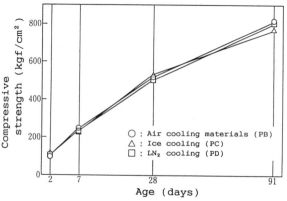

Fig 3. Cooling method and strength development

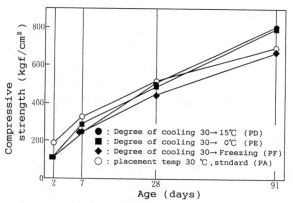

Fig 4. Cooling degree and strength development

3.2.2 Pore volume and pore distribution

The mercury pressurized injection method was used to measure the pore volume and pore distribution. The measurment sample was obtained by pulverizing a test piece immediately after a strength experiment, selecting samples between 2.5 and 5 mm in size, terminating hydration with acetone, and vacuum drying them before performing the measurements.

The pore structure was studied over a range of 43 to 75,000 angstroms, and the total pore volume was obtained using the total volume of pores in this range. The pore distribution was organized within the range of 43 to 3200 angstroms.

Figure 5 shows the results of regression analysis of the strength and the pore volume of the cement paste. It reveals an extremely strong correlation between strength development and change in total

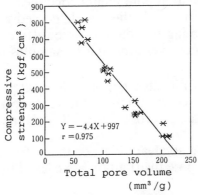

Fig 5. Regression analysis of
the strength and the pore volume

pore volume. These results clearly demonstrate that the strength increases as the pores in the cement paste decline along with progressive hydration.

Table 4 shows the relationship between the total pore volume and cooling methods using various cooling media. Figure 6 presents, as an example, a comparison of pore distribution in a sample produced by cooling with liquid nitrogen (PD) and one produced by cooling the materials with air prior to mixing (PB).

The table demonstrates that no matter which cooling medium is used, as the ages, the total pore volume changes in almost exactly the same manner. The figure shows that changes in pore distribution that occur as the ages can be organaized as follows.

During initial aging, large diameter pores are found, and they are uniformly distributed between 43 to 3200 angstroms. As aging progresses, however, the pore distribution shifts toward smaller

Table 4. Difference in cooling method and change in total pore volume (mm³/g)

Age	Experiment cord		
	PB	PC	PD
2 days	218	209	213
1 week	159	156	156
4 weeks	101	104	102
13 weeks	63	63	57

Table 5. Difference in cooling degree and change in total pore volume (mm³/g)

Age	Experiment cord			
	PA	PD	PE	PF
2 days	204	213	203	212
1 week	156	156	138	165
4 weeks	113	102	110	109
13 weeks	74	57	56	63

diameter pores, and the total pore volume declines.

In addition, the distribution in the sample cooled with ice (PC)is also similar, demonstrating that there is little difference in the pore distribution changes over time among samples using different cooling methods.

Table 5 shows the relationship between the degree of cooling and the total pore volume, wheras Figure 7 presents a comparison of the pore distribution in a sample cooled to 15°C (PD) and one that was over–cooled (PF).

These results provide no clear evidence that differences in the total pore volume and pore distribution are caused by the degree of cooling performed.

3.2.3 Extent of hydration

The degree that hydration of cement has progressed can be assessed according to the quantity of combined water.

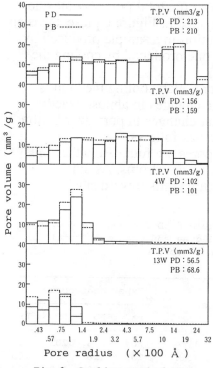

Fig 6. Cooling method and
pore distribution

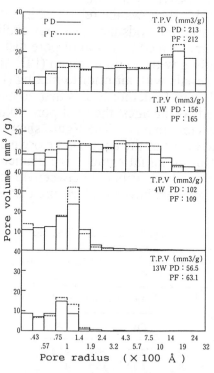

Fig 7. Cooling degree and
pore distribution

96

Various methods of measuring the quantity of combined water are in use, but in order to compare differences among cooling methods and differrences caused by the extent of cooling, the quantities of combined water were obtained as pro-rate of released water quantity up to 950°C using thermal analysis equipment. Table 6 shows the quantity of combined water of the specimen after 13 weeks of aging.

As the table clearly indicates, the quantity of combined waters in PB and PD which were pre-cooled to the same degree by different cooling methods are very similar.

Turning to the degree of cooling performed with liquid nitrogen, it is possible to conclude that despite the fact that the combined water content detected when the material was cooled to 0°C (PE) was slightly lower than when it was only cooled to 15°C (PD), the proportion of combined water could be considered to be almost identical. When a sample was overcooled to the extent that it was frozen for a time, however, the volume of combined water tends to be a little lower.

This enables to conclude that even when liquid nitrogen is employed as a cooling agent, it has little effect on the progress of the hydration of the cement. It also demonstrates that partial over-cooling has little adverse effect on hydration.

3.2.4 Microstructure of the hardened cement paste

Figure 8 and 9 present SEM images of fractures in a sample of cement paste cooled to a casting temperature of 30°C (PA) and a sample cooled to a casting temperature of 15°C (PD) with liquid nitrogen after 1, 4, and 13 weeks of aging. The SEM images were selected at random.

The SEM images at a casting temperature of 30°C without cooling (Fig 8) show that after 1 week of aging, surface unevenness is conspicuous and many gaps exist. SEM images made after 4 weeks and 13 weeks, however, reveal a tendency for the extent of the surface

Table 6. Combined water content measurement results (13 weeks)

Experiment cord	PB	PD	PE	PF
Combined water(%)	22.2	22.4	21.7	20.8

Measurement condition	Atmosphere : Air
	Samples : 25 to 30 mg
	Temperature increase rate : 10°C /min
	TG full scale : 5 mg

irregularities to decline as the ages. After 13 weeks, when the cement paste becomes extremely strong, the surface area is very smooth, and the micro–structure has become particularly fine. The results of these observations are identical for the SEM images of sample cooled with liquid nitrogen (Fig 9). Based on this, it can be

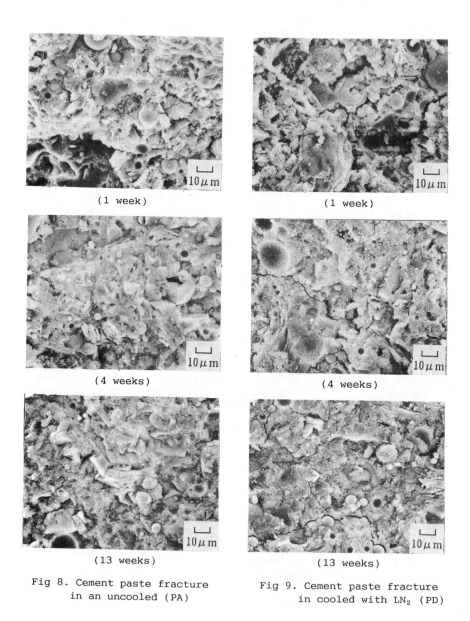

(1 week)   (1 week)

(4 weeks)   (4 weeks)

(13 weeks)   (13 weeks)

Fig 8. Cement paste fracture   Fig 9. Cement paste fracture
    in an uncooled (PA)   in cooled with LN₂ (PD)

concluded that even cooling with liquid nitrogen has no effect on the cement paste microstructure.

Next the structure of cement paste produced using various cooling method and cooled to various degree were compared. Detailed results are not presented here, but the results of observations of all samples are similar to the above mentioned results. No clear differences were observed in the structures of the cement paste.

4 Conclusions

The following points sum up the results of experiments on concrete and cement paste cooled with liquid nitrogen.

1) The experiments verified that it is possible to improve strength development by pre–coolineg high–strength mass concrete with liquid nitrogen.

2) The experiments demonstrated that there is no difference in the strength and microstructure of various cement paste samples that were cooled to the same degree with various cooling media.

3) Even when concrete was cooled to 0 degrees centigrade with liquid nitrogen, the strength development of the concrete or the cement paste was unaffected, and the cooling had no adverse effect on the microstructure of cement paste.

5 Reference

1) Inoue, K., Nakane, S., Ohike, T., et.al, (1985)
   Experimental Study on Quality Control of High Strength Mass Concrete for PCCV (Phase 1 to 7), Summaries of Technical Papers of Annual Meeting Architectural Institute of Japan (Tokai), October, pp 313–326.
2) Sakamoto, T., Nakane, S., Kawaguchi, T., et.al, (1987)
   Study on Concrete Strength Development in High Strength Massive Structure (Phase 1 to 7), Summaries of Technical Papers of Annual Meeting Architectural Institute of Japan (Kinki), October, pp 221–234.
3) Nakane, S., Sogo, S., et. al, (1986)
   Basic Properties of Concrete Cooled With Liquid Nitrogen, Proceedings of the Japan Concrete Institute Vol 8. pp 329–332.

10 THE INFLUENCE OF HIGH TEMPERATURE ON THE STRENGTH AND PORE STRUCTURE OF CONCRETES MADE WITH A NATURAL POZZOLAN

J. G. CABRERA
Department of Civil Engineering, University of Leeds, UK
S. O. NWAUBANI
Department of Civil Engineering, University of Surrey,
Guildford, UK

Abstract
This paper present the results of a study carried out to evaluate the effect of curing temperature on the strength and pore-structure of concrete and mortar containing pozzolanic additions. The natural pozzolan used is a red tropical soil (RTS) which is present in the B and C horizon of deep chemically weathered profiles typical of tropical and subtropical regions. Laboratory specimens were cured at 20°C and 45°C. The results are compared with those obtained for similar concretes and mortars made with ordinary portland cement and with addition of pulverised fuel ash (pfa).
Keywords: Concrete, Pozzolans, Hot Climates, Porosity, Pore Structure, Strength.

1 Introduction

The use of mineral additions for the manufacture of concrete include generally natural pozzolans from volcanic origin, artificial pozzolans like pulverised fuel ash and microsilica and other hydraulic binders like ground granulated blast furnace slag. Natural pozzolans produced by intensive chemical weathering in tropical regions have not been used extensively. Murat (1983), Ambroise et al (1986) and Cabrera & Nwaubani (1992a) have reported its use as a mineral addition to cement in concrete, however information regarding its effects on the long term performance of concrete is limited. The use of these pozzolans known generically as red tropical soils (RTS) (Cabrera and Nwakanma, 1980) is potentially very attractive since they are very abundant in tropical and subtropical regions.

Studies on the lime reactivity of RTS have shown that the silica, alumina and iron oxide phases react at normal temperature with $Ca(OH)_2$ to form solid solutions of calcium silicate hydrates, calcium aluminates hydrates, hydrated gehlenite and calcium aluminoferrites and that, as with many other natural pozzolans, thermal treatment can increase their lime reactivity (Murat (1983), Cabrera and Nwakanma (1980), Nwaubani (1990)). For this reason it is apparent that its use in concrete can be of benefit in terms of performance and cost.

This paper presents the results of measurements of strength, porosity and pore size distribution carried out in specimens cured at 20° C and 45°C for periods up to one

Concrete in Hot Climates. Edited by M. J. Walker. © RILEM
Published by E & F N Spon, 2 - 6 Boundary Row, London SE1 8HN. ISBN 0 419 18090 7.

year. Two RTS natural pozzolans and the same pozzolans activated at 800°C for two hours were used as additions to opc and the concretes and mortars were compared with mixtures made with opc or with opc plus pfa.

Total porosity and average pore diameter as a numerical parameter to describe pore structure are used in conjunction with compressive strength to characterize the performance properties of the concretes and mortars studied.

2 Materials

2.1 Red tropical soils (RTS)

Two red tropical soils from the Southern Indian Shield of Sri Lanka were obtained for use in the experiments. The preparation of these materials consisted of pre-drying at room temperature, crushing, wet sieving through a 75 micron sieve and final drying at 60°C to avoid the risk of modifying the amorphous iron oxides contained in the natural soils.

The thermal activation of these soils consisted in heating them in a furnace at 800°C for two hours. This temperature and residence time resulted in the collapse of the poorly ordered kaolinite clay mineral which converted to the amorphous meta-kaolinite and thus increased its lime reactivity as shown by Cabrera and Nwaubani (1992).

2.2 Pulverised fuel ash (pfa)

The pulverised fuel ash used was supplied by National Power as an unclassified material originated from Drax Power Station in the North of England.

2.3 Ordinary portland cement (opc)

The ordinary portland cement used was supplied by Castle UK Ltd.

2.4 Aggregates

Quartzitic sand and gravel were used for the preparation of mortars and concretes. These were obtained from commercial deposits in the County of Nottinghamshire, England. The sand gradation conformed to the requirements of zone M or the British Standard and the gravel was nearly single sized with maximum size of 14 mm also conforming to the British Standard BS882 (1983).

2.5 Superplasticiser

The superplasticiser used was napthalene formaldehyde condensate supplied by FEB UK.

Table 1 shows the main oxides and other physical and mineralogical properties of the cement and the pozzolans used.

Table 1 Chemical and physical properties of the materials

Properties Oxide Composition (%)	OPC	PFA	RTS1/ARTS1	RTS2/ARTS2
SiO_2	20.40	50.80	39.06	42.40
Al_2O_3	5.02	27.90	31.65	35.14
Fe_2O_3	2.92	11.70	12.89	3.61
MnO	0.06	-	0.07	0.02
TiO_2	0.21	1.00	0.83	1.67
CaO	64.25	1.20	0.09	0.11
MgO	2.83	1.50	0.52	0.41
Na_2O	0.39	0.80	0.27	0.17
K_2O	0.84	3.70	0.44	0.36
P_2O_5	0.08	-	0.18	0.09
SO_3	2.63	0.59	-	-
LOI	0.70	2.10	15.95	16.67
Specific gravity	3.15	2.24	2.80	2.66
Spec surface, m^2/g (BET)	1.04	1.29	57.97*	44.07*
Spec surface, m^2/g (BET)	-	-	44.30**	37.80**
Plasticity Index	-	-	18.00	35.00
Amorphous Ferri-Alumino-Silicate (%)	-	-	12.38	9.04
Alkali soluble Silica + Alumina	-	36.17	-	-

*   RTS1 & RTS2 = Natural red tropical soils
** ARTS1 & ARTS2 = Activated RTS at 800^0 for 2 hours

3 Preparation of specimens and curing

The concrete mix was designed according to the method specified in BS5328 (1981). The cement or cement plus pozzolan content was 360 kg/m^3. The specimens containing RTS were made with variable quantities of superplasticiser in order to maintain equal workability to the opc control mix. The workability of the different mixes gave slump values of 55 ± 5 mm and compacting factors of 0.92 ± -2. The mix containing pfa did not require reduction of water to maintain the same range of workabilities, as it is usually the case when using pfa. This is because the percentage replacement was only 15% for pfa and RTS.

The composition of the concrete mix in terms of cement sand and gravel was 1:2.04:3.07. The pozzolan replacement was 15% of the weight of cement. The w/c ratio was maintained constant at 0.5 varying the quantity of superplasticiser used to maintain constant slump of 50 ± 5 mm as shown in Table 2.

The mortar specimens were prepared with the same proportions used to make the concrete mixes but without the gravel. The dosage of superplasticiser required to achieve equal consistency between the pozzolanic mortars and the opc mortar was slightly lower than the dosage used for the concrete mixes. The workability of the mortars was controlled using a standard mortar flow-table as indicated in BS4551 (1980). Table 2 shows the measured flow-table spreads obtained for the various mixes at the indicated dosages of superplasticiser.

The concrete specimens were 100 x 100 mm cubes, while the mortar specimens were cylinders of 100 mm diameter and 25 mm thickness. Both sets of specimens were demoulded after 24 hours and placed under water at $20^{0}C$ or $45^{0}C$ until the time of testing. The curing periods were 3, 7, 28, 90, 180 and 360 days for the concrete specimens while for the mortar specimens the maximum curing time was 90 days.

The code used for the different mixes was as follows:

opc   = control mix made with ordinary portland cement
pfa   = mix made with 0.85 opc + 0.15 pulverised fuel ash
RTS1   = mix made with 0.85 opc + 0.15 natural red tropical soil type 1
ARTS1 = mix made with 0.85 opc + 0.15 activated red tropical soil type 1
RTS2   = mix made with 0.85 opc + 0.15 natural red tropical soil type 2
ARTS2 = mix made with 0.85 opc + 0.15 activated red tropical soil type 2

Table 2. Superplasticiser dosages for concrete and for mortar mixes and flow-table spreads for mortar mixes

Mix	Concrete	Mortar	F-T spread (mm)
	Superplasticiser	(%)	
opc	0.00	0.00	199.9
pfa	0.00	0.00	202.1
RTS1	0.70	0.60	202.5
RTS2	0.90	0.80	200.0
ARTS1	0.44	0.30	201.3
ARTS2	0.53	0.42	201.9

4 Methods of testing

4.1 Compressive strength

The concrete specimens were used to obtain values of compressive strength. The test followed the procedure outlined in BS1881 (1983).

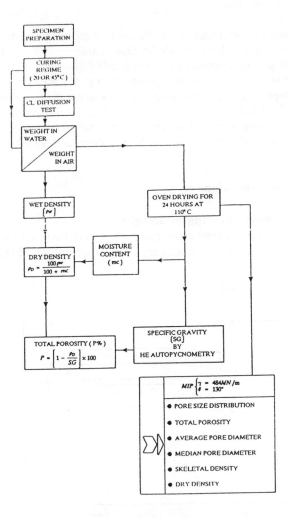

Fig.1. Flow chart of the procedure used for determination of
total porosity and pore size distribution

4.2 Total porosity

Total porosity values were obtained using the mortar specimens. Helium pycnometry was used to determine the specific gravity (relative density) of the mortar mixes and this value in conjunction with the dry density of the mortar allowed to calculate the total porosity (Cabrera (1985)). Figure 1 shows the flowchart of the procedure used to calculate porosity and it includes the appropriate numerical relations necessary for this calculation.

Total porosity values were also determined while obtaining the pore size distribution by mercury intrusion porosimetry. These values will be compared with the values obtained by helium pycnometry.

4.3 Pore size distribution

The pore size distribution measurements were carried out using mercury intrusion porosimetry with a Micromeritic Autopore Model 9200 instrument. The mercury contact angle used 130⁰ and the value of surface tension 484 dynes/cm. The mercury porosimeter is capable of generating intrusion pressures up to 414 MPa and therefore measure pore size diameters up to 0.003 micron. Figure 1 shows the procedure of sample preparation and a list of the parameters obtained from the mercury intrusion test. Chloride diffusion values were also measured as is shown in Figure 1. However the results are not reported in this paper but elsewhere (Cabrera and Nwaubani (1992b)).

5 Presentation and discussion of the results

5.1 Influence of temperature of curing on strength

Figure 2 presents the results of concrete compressive strength of the mixes cured at 20⁰C in water. Table 3 shows the same results expressed as percentages of the opc control mix.

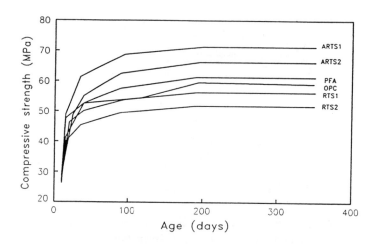

Fig.2. Compressive strength - age relationships for concretes
cured at 20⁰C in water

Table 3. Compressive strength results of concrete mixes cured at 20°C.

Mix	Strength expressed as a percentage of the opc mix Curing age (days)			
	3	28	90	365
opc	100	100	100	100
pfa	81	103	102	103
RTS1	90	103	99	97
RTS2	87	92	91	89
ARTS1	114	120	115	114
ARTS2	105	107	110	108

At 3 days of age the activated pozzolans gave higher strengths than the opc control mix, above the values expected due to normal statistical laboratory variation ($\pm$ 4 MPa). The natural pozzolans and the pfa gave lower values of strength than the opc mix. These results are in line with the lime reactivity of the pozzolans (Cabrera and Nwaubani (1992c)). The activated RTS have a distinctly higher reactivity than pfa or the natural materials. It is interesting to note that at constant w/c ratio the 28 day strength of the pfa and RTS1 mixes have reached strength values equal to the opc control. The trend is maintained up to one year of curing. With exception of the RTS2 natural pozzolan mix all others perform satisfactorily from the point of view strength requirements. The activated pozzolans show a distinct improvement from very early curing age.

For specimens cured at high temperature in order to simulate a mean tropical temperature (45°C in water) the trends are different. Figure 3 shows the strength development with age while Table 4 shows again the strength values expressed as percentages of the opc control mix.

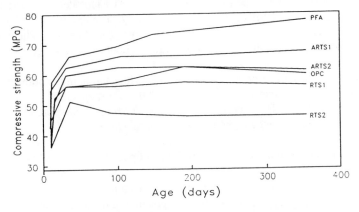

Fig.3. Compressive strength - age relationships for concretes
cured at 45°C in water

The pfa mix shows a very different development of strength. At three days of age the strength of the pfa mix is 89% of the opc mix but by 28 days the strength has reached 117% and at one year this difference is increased to 124%. This trend confirms other findings which show that at ages of 25 to 30 years the percentage difference between opc and pfa mixes is maintained (Cabrera and Woolley (1985)).

Table 4. Compressive strength results of the concrete mixes cured at 45°C

Mix	Strength expressed as a percentage of the opc mix Curing age (days)			
	3	28	90	365
opc	100	100	100	100
pfa	89	117	117	124
RTS1	103	100	99	95
RTS2	87	92	82	75
ARTS1	113	109	110	107
ARTS2	107	106	106	100

The activated RTS pozzolans show that at 3 days of age the strength is 7 to 13% higher than the control mix, however the difference in strength decreases as the curing age increases. This decrease in strength with age is related to the crystallographic changes which occur with the calcium aluminate hydrates at high temperature as shown by Nwaubani (1990). The conversion from the hexagonal to the cubic form of the hydrates increases the porosity of the concrete and thus reduces its strength.

It is also interesting to point out that the strength of the opc mix cured either at 20 or 45°C after one year is approximately the same (see Figures 2 and 3)) while the strength of the pfa mix has increased from 68 to 77 MPa. The activated ARTS1 and ARTS2 show a decrease of 11 MPa and 8MPa respectively. The loss of strength of the natural RTS2 pozzolan is still higher. At 20°C and one year of age the value was 58.5 MPa while at the same age but at high temperature of curing the strength reduced to 46.5 MPa.

Natural pozzolans reach in alumina should be used with caution in hot climates. The same materials when activated for two hours at 800°C show superior performance to concretes made with ordinary portland cement and therefore it is clear that its use should be recommended when they have been activated.

5.2 The influence of curing temperature on the total porosity and total porosity-strength relationships

Although the measurements of porosity were obtained from mortar specimens, knowing the porosity of the aggregate and the composition of the mix it was possible to calculate an acceptable approximate value of total porosity for the concrete mixes. The results show that as the mixes hydrate their porosity values decrease and that there is a good correlation between porosity and strength. For every mix there is a distinct numerical relationship. Table 5 gives the equations relating strength to porosity for both curing regimes. The equations show clearly that for a constant strength, different mixes have different porosities. This is an indication that the pore size distribution for each mix must be different and therefore to obtain a correlation which is valid for all mixes there is need to incorporate a numerical parameter which describes the pore size distribution. The relations between strength and porosity shown in Table 5 give a good indication of the durability characteristics of the mixes studied. For example a 50 MPa concrete made with the ARTS1 pozzolan and cured at 20°C has a total porosity of 9.2% while a concrete of the same strength made with opc has a porosity of 11.8%. This is a considerable difference in available pore volume for permeability or diffusion resistance.

Table 5. Equations showing the relationship between compressive strength and total porosity of the concrete mixes

Mix	Equation	r^2	Level of Signific. PR>F
	Curing temperature = 20°C		
opc	CS = 208.38 - 64.16 ln P	0.974	0.0053
pfa	CS = 213.74 - 67.42 ln P	0.982	0.0030
RTS1	CS = 216.30 - 68.75 ln P	0.950	0.0132
RTS2	CS = 197.20 - 60.54 ln P	0.988	0.0020
ARTS1	CS = 271.30 - 99.72 ln P	0.995	0.0005
ARTS2	CS = 213.23 - 78.90 ln P	0.971	0.0060
	Curing temperature = 45°C		
opc	CS = 135.97 - 34.90 ln P	0.926	0.0018
pfa	CS = 197.97 - 63.16 ln P	0.974	0.0010
RTS1	CS = 175.47 - 52.27 ln P	0.934	0.0064
RTS2	CS = 178.99 - 55.41 ln P	0.926	0.0079
ARTS1	CS = 180.97 - 58.53 ln P	0.994	0.0001
ARTS2	CS = 176.57 - 54.60 ln P	0.938	0.0056

CS	=	compressive strength (MPa)
ln P	=	natural logarithm of porosity (%)
r^2	=	correlation coefficient
PR>F	=	level of significance

The total porosity of the mortars measured using helium pycnometry and the values obtained while measuring pore size distribution by mercury intrusion porosimetry (mip) gave comparable results. The values obtained by mip were marginally lower when measuring high porosity samples. This is probably because the helium gas is able to gain access into most of the finer pores which the larger molecules of mercury cannot enter (Beaudoin (1979)).

The statistical relation ($r^2 = 0.821$) between both porosities gave the following equation:

$$Ph = 1.19\,Pm - 4.21 \tag{1}$$

Ph = Helium porosity (%)
Pm = Mercury intrusion porosity (%)

5.3 Development of a model to quantify the relation between compressive strength and pore structure of concrete

The pore size distribution results showed that the capillary pore volume decreases with time of curing and that the pore size distribution shows an increase in the very fine pores. The clear shift towards fine pore structure can be quantified by expressing the pore size distribution in terms of the average pore diameter. An exception to these trends was found with the natural pozzolans, mainly with the RTS2 pozzolan mix which at 45°C and 90 days of age showed a coarsening of the pore structure. This confirms the view that at high temperature there is a crystallographic change of the aluminates which results not only on a higher pore volume but also a coarser pore size distribution.

An example of the pore size distribution change with age is shown in Figure 4. This corresponds to a pfa mix cured at 45°C. The total porosity of this mix reduced from 22% at three days to 14.75% at 90 days and its average pore diameter reduced from 0.057 micron to 0.031 micron.

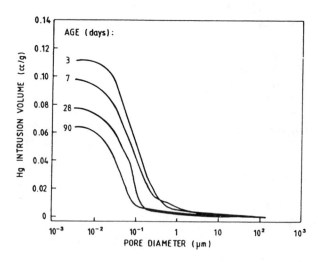

Fig.4. Changes in pore size distribution of the pfa mortar
mix cured at 45°C

Table 6 gives a summary of the mip results in terms of mean pore diameters and total porosities for all the mortar mixes studied at 3 days and 90 days of age.

Using the values of average pore diameter in conjunction with the values of concrete total porosity it is possible to find a valid statistical equation which describes the relation between compressive strength and pore structure characteristics for concretes cured at normal and high temperature. The relationship found is shown in Figure 5. The numerical expression which gave an acceptable correlation coefficient of 0.73 is of the following form:

$$CS = 4.797 \ (1/d^2 + P) + 5.41 \tag{2}$$

where:

CS = concrete compressive strength (MPa)

d = average pore diameter (micron)

P = porosity of concrete as a fraction

Table 6. Mip total porosity and average pore diameter values for the mortar mixes studied

Mix	Age (days)	Total porosity (%)		Average pore diameter (micron)	
		20 degree C	45 degree C	20 degree C	45 degree C
opc	3	23.14	20.77	0.0707	0.0601
	90	17.31	15.56	0.0804	0.0308
pfa	3	24.03	21.90	0.0569	0.0570
	90	16.73	14.75	0.0508	0.0310
RTS1	3	23.29	19.73	0.0571	0.0490
	90	16.93	17.18	0.0342	0.0190
RTS2	3	26.96	21.96	0.0470	0.0469
	90	17.94	18.93	0.0330	0.0360
ARTS1	3	20.63	19.69	0.0546	0.0302
	90	14.07	14.70	0.0370	0.0228
ARTS2	3	20.66	19.72	0.0599	0.0487
	90	14.63	15.49	0.0324	0.0300

To make this statistical model useful as a laboratory guide for the design of concretes of adequate performance it is necessary to determine minimum acceptable values of the parameter $1/(d^2 + P)$ which describes the characteristics of the pore structure of concrete. The determination of minimum values of the pore structure parameter has to be done in relation to the nature of the environment where the concrete will be used. Tentative values for very aggressive, aggressive and mild environments are proposed elsewhere (Cabrera and Nwaubani (1992d)).

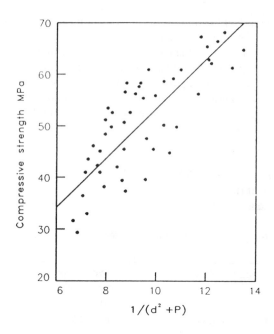

Fig.5. Model which relates compressive strength of all concretes
with a parameter describing pore structure

6 Conclusions

From the data presented in this paper the following conclusions are offered:

1. Activated RTS pozzolan concrete mixes exhibit higher strengths than the equivalent opc concrete mix when cured at 20°C. The pozzolanic mixes gave strengths which were 5% to 14% greater than the opc mix strength and this percentage did not change with age of curing. The natural pozzolan mixtures gave values slightly lower than the values of the control mix.

2. Curing at 45°C resulted in early high strengths for the pozzolanic mixes. With increasing curing time the difference between the ARTS mixes and the opc control decreased. The mixes made with the natural RTS showed a decrease in strength with time. This has been attributed to the crystallographic changes which take place when metastable aluminates are exposed to high temperature and high moisture environments. In view of these results, natural pozzolans containing appreciable quantities of alumina should only be used in hot environments when they have been activated thermally.

3. An exponential decay function which describes satisfactorily the relation between compressive strength and porosity was found for each mix, however a general function to relate strength to porosity for all mixes does not exist. This is an

indication that it is not only total porosity which is important but some other parameter to describe the pore structure of the different mixes.

4. A statistical model relating compressive strength to pore structure is proposed. Pore structure is numerically described by the parameter $1/(d^2 + P)$. Limits for this parameter should be established as a function of concrete exposure environmental conditions.

7 References

Ambroise, J.C. Murat, M. and Pera, J. (1985) Hydration reaction and hardening of calcinated clays and related minerals, V Extension of the research and general conclusions. **Cem & Conc. Res.,** 15, 261-268.

Beaudoin, J.J. (1979) Porosity measurements of some hydrated cementitious systems by high pressure mercury intrusion - microstructural limitations. **Cem & Con Res.,** 9, 771-781.

British Standard Institution. (1980) Standard methods for testing mortars, screeds and plaster. **BS4551:1980,** BSI, London.

British Standard Institution (1981) The mix design process. **BS5328:Part2:1981,** BSI, London.

British Standard Institution. (1983) Coarse and fine aggregates from natural sources for concrete. **BS882:Part2:1983,** BSI, London.

British Standard Institution. (1983) Methods for determination of compressive strength of concrete cubes. **BS1881:Part 116:1983,** BSI, London.

Cabrera, J.G. (1985) The porosity of concrete. **Con Soc Res Sem.,** Leeds.

Cabrera, J.G. and Nwakanma, C.A. (1980) Properties of red tropical soils treated with lime. **Proc. 7th Regional Conf. for Africa. Soil Mech. & Found. Eng.,** 2, 18-32.

Cabrera, J.G. and Nwaubani, S.O. (1992a) Strength and chloride permeability of pozzolanic mortars and concretes. Offered to **Magazine of Concrete Research.**

Cabrera, J.G. and Nwaubani, S.O. (1992b) The microstructure of concrete made with pozzolanic additions and their resistance to ionic ingress. **IAHS Int. Cong. on Housing.** Birmingham.

Cabrera, J.G. and Nwaubani, S.O. (1992c) Lime reactivity of red tropical soils. Offered for publication to **Cement & Concrete Composites.**

Cabrera, J.G. and Nwaubani, S.O. (1992d) The assessment of performance of pozzolanic concrete using a pore structure parameter. (In preparation).

Cabrera, J.G. and Woolley, G.R. (1985) A study of twenty five year old pfa concrete used in foundation structures. **Proc. Inst of Civil Engineers,** 2, 79, 149-165.

Murat, M. (1983) Hydration reaction and hardening of calcinated clays and related minerals. 1. Preliminary investigation on metakaolinite. **Cem & Conc. Res.,** 13, 2, 259-266.

Nwaubani, S.O. (1990) **Properties, Hydration and Durability of Pozzolanic Mortars and Concretes.** Unpublished PhD Thesis, University of Leeds.

11 PERFORMANCE PROPERTIES OF POZZOLANIC MORTARS CURED IN HOT DRY ENVIRONMENTS

P. J. WAINWRIGHT, J. G. CABRERA and A. M. ALAMRI
Department of Civil Engineering, University of Leeds, UK

Abstract
The performace properties of mortars made with ordinary portland cement (opc) and pozzolanic cements containing either pulverised fuel ash (pfa) or ground granulated blast furnace slag (ggbs) have been measured after exposing the mixes to laboratory simulated hot dry environments.
The simulated environments were: $20^0C + 70\%$ RH, $350^0C + 70\%$ RH and $45^0C + 30\%$ RH. Specimens were cured for different lengths of time before testing. The tests carried out to assess the performance properties and thus the durability of the mortars were: total porosity, pore size distribution and gas permeability using oxygen. The tests showed that performance of the mortar mixes is enhanced by increased curing time. Uncured specimens subjected to hot dry environments ($45^0C + 30\%$ RH) were strongly affected in their durability characteristics as shown by the deterioration of the performance indicators. Opc mortar were severely affected by the hot dry environments independently of the length of curing. Pozzolanic mortars subjected to curing periods of one day or more in hot dry environments exhibited better properties than equivalent mortars cured at normal temperatures.
Keywords: Mortar, pozzolan, Hot climate, porosity, pore structure, permeability, pulverised fuel ash, blast furnace slag.

1 Introduction

The long term performance of concrete in a structure is a function more of durability than strength. Durability is an attribute of concrete which is related to its ability to resist attack from the environment in which it is placed, to maintain its appearance and to continue to function in the manner for which it was designed. Unlike strength, durability is not easy to define or to measure. The ability of the concrete to resist ingress of any deleterious material is a good indication of its durability but it is unlikely that this can be linked to one single parameter. Porosity and permeability to air, water, water vapour, chlorides and sulphates are probably the more important parameters considered to have an influence on durability.

Most engineers agree that concrete must be well cured in order for it to achieve adequate performance and that curing is even more important in hot dry environments

Concrete in Hot Climates. Edited by M. J. Walker. © RILEM
Published by E & F N Spon, 2 - 6 Boundary Row, London SE1 8HN. ISBN 0 419 18090 7.

than in cooler damp environments. There is a lack of experimental evidence available regarding the influence of curing on the durability of concrete. Senbetta and Scholer (1984) carried out tests to measure the water absorption of concrete at various depths below the surface and concluded that curing only influences the top few centimetres of the concrete. This view was confirmed by Wainwright et al (1985) Cabrera et al (1989) and Wainwright and Cabrera (1990) in tests they carried out to measure the oxygen permeability of concrete at various depths beneath the surface of slabs cured using different types of curing membranes. The studies were made using ordinary Portland cement concrete. The advantages of using blended cements containing either pulverised fuel ash (pfa) or ground granulated blastfurnace slag (ggbs) has now been accepted particularly in Europe, Cabrera (1985), Reeves (1985), but because both materials are slower to react than the opc which they replace there are some doubts as to their performance when they are inadequately cured and particularly so in hot dry environments. The general opinion is that in such situations the blended cements will perform worse than opc but there is little quantifiable evidence to confirm this belief.

This study reports the results of a laboratory investigation carried out on mortar specimens kept under three different environmental conditions and cured for different lengths of time. The mortars were made with opc and with blends of opc/pfa and opc/ggbs. The durability was assessed by measuring oxygen permeability, porosity, and pore size distribution; all tests were carried out at the age of 28 days.

2 Experimental details

2.1 Materials
The sand used was quartzite in origin conforming to the zone M requirements of BS882 (1983). The pfa was supplied by the Central Electricity Generating Board from Drax Power station in the north of England and the slag was supplied by the Frodingham Cement Company, Scunthorpe. The cement used was ordinary Portland (opc) and the composition of all materials is given in Table 1.

2.2 Curing and environmental conditions
Three different curing environments were chosen, two to simulate Middle Eastern conditions and one less aggressive more temperate environment; details of these are given in Table 3. In an attempt to simulate as closely as possible both site conditions and practice, materials were stored for one week prior to casting in the environment in which the respective mortar specimens were to be kept. Due to the size limitation of the environmental room the actual casting had to be carried out in the laboratory at a nominal temperature of 18°C. Immediately after casting all specimens were returned to the controlled environment and the initial mix temperatures achieved at this stage were recorded and are shown in Table 4.

Table 1 Composition of the cementitious materials used

Constituent	Percent by weight		
	opc	pfa	ggbs
SiO_2	20.9	51.9	36.8
Al_2O_3	5.5	26.9	10.0
Fe_2O_3	2.7	11.3	1.2
CaO	64.3	1.5	41.9
MgO	2.5	1.6	7.2
Na_2O	0.3	1.2	0.3
K_2O	0.8	3.8	0.5
TiO_2	-	0.9	0.6
SO_3	2.8	0.6	0.1
Alkali-Soluble SiO_2	-	30.0	-
Alkali-Soluble Al_2O_3	-	6.2	-
Loss on ignition	0.7	2.6	2.3
Specific Surface m²/g	0.38	0.22	0.40

2.3 Mix proportions
Details of the mix proportions used are given in Table 2.

Table 2 Mortar mix proportions (by mass)

mix	opc	pfa	ggbs	sand	water
1	1	0	0	2.3	0.45
2	0.7	0.3	0	2.3	0.45
3	0.4	0	0.6	2.3	0.45

Table 3 Details of curing environments

Environment	Description
1	$20^0C \pm 2^0$ and 70%RH $\pm$ 5%RH
2	$35^0C \pm 2^0$ and 70%RH $\pm$ 5% RH
3	$45^0C \pm 2^0$ and 30% RH $\pm$ 5% RH

Table 4 The Initial mix temperatures of the mortar mixes

mix	environment	
	$35^0C + 70\%RH$	$45^0C + 30\%RH$
1	29^0C	38^0C
2	30^0C	37^0C
3	29^0C	37^0C

All the specimens cast were 50 mm cubes, these were demoulded at an age of one day, all sides, except the top as-cast face, were then covered and sealed using polythene sheet. With the exception of those that were to be uncured, the top as-cast face of the remaining specimens were covered immediately after casting with polythene sheet. Wet hessian was applied at approximately six hours and this was kept moist and covered with polythene for the duration of the curing period. Specimens were cured for one of five different lengths of time namely:- uncured, one-day, three-day, seven-day and continuously cured for 28 days. After curing, the specimens remained in the controlled environment until testing, all testing was carried out at the same age of 28 days.

2.4 Tests carried out
The tests used to evaluate the influence of curing in the three different environments were:

2.4.1 Oxygen Permeability

This test was performed using the equipment and procedure developed by Cabrera and Lynsdale (1988). Specimens used were 25.4 mm in diameter by 48-50 mm in height cored from the 50 mm cubes.

The samples to be tested were removed from their respective curing environments at the age of 24 days, cured and placed in an oven at 105°C $\pm$ 5° for 72 hours. They were then placed in an air-tight container until they reached room temperature before the start of the test.

2.4.2 Pore size distribution

The pore size distribution of the mortar specimens was examined using a Micrometirics Autopore 9200 mercury intrusion porosimeter (mip). The samples tested were 25.4 mm diameter discs 12-14 mm thick taken 10 mm below the top surface of cores cut from the cubes. Specimens were taken at an age of 24 days and conditioned in the same way as those tested for permeability.

2.4.3 Porosity

The porosity of a mortar is calculated following the method proposed by Cabrera (1985). Briefly it consists of measuring the wet density of the mortar, correcting for moisture content and calculating the dry density. This density is used in the following equation:

$$Pm = 1 - \frac{Dd}{SG} 100 \qquad (1)$$

where: Pm = porosity of mortar, percent
           Dd = dry density g/cc
           SG = specific gravity

The specific gravity is measured by helium pycnometry. The porosity then corresponds to the volume of pores which can be penetrated by helium when the sample has been ground to a fineness where the particles are smaller than 75µm. The porosity of the sand used for the mortars is measured following the same method. The difference of Pm and the porosity of the sand is the porosity of the hydrated paste.

3 Presentation of results

3.1 Oxygen permeability

The coefficient of permeability results are shown in Fig.1. A statistical analysis carried out on the differences between readings from nominally identical specimens has indicated a degree of repeatability of $\pm$ 10% at 95% confidence limits.

Figure 1. Relation between coefficient of permeability and curing time

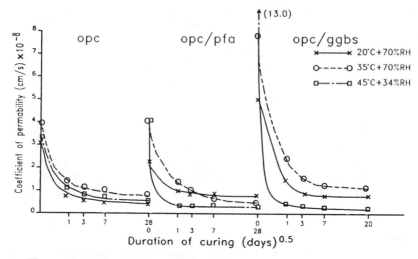

The results show that permeability decreases with increasing curing time. The largest reductions taking place between specimens that were not cured and those cured for just one day. Compared with those specimens which were not cured, specimens cured for one day were at least 50% less permeable. The most dramatic reduction was observed with those opc/ggbs samples kept at 40°C and 30% RH, the coefficient of permeability was seen to reduce from 13 x 10^-8 to 0.45 x 10^-8 cm/sec. Increasing the curing period beyond one day produced further but less significant reductions in permeability and the differences between the 7 day and the continuously cured specimens was only marginal.

For the uncured specimens the least aggressive of all the environments was 20°C and 70% RH and, with the exception of the opc/ggbs samples, there was little difference between the two hotter environments. In all cases, for curing periods of one day and more the most aggressive environment would appear to be that of 35°C and 70% RH.

Comparing the three different mixes used, that containing ggbs was far more sensitive to lack of curing in all environments than the other two mixes both of which showed similar trends under uncured conditions. There is no doubt that both the pfa and the slag mixes perform far better than the opc mix for curing periods of one day or more in the hottest environment of 40°C and 30% RH although the differences in the other two environments are less clear.

3.2 Pore structure

The pore structure of a solid is generally characterised by it total porosity and pore size distribution, however pore size distribution functions are difficult to use as performance indicators. It has been shown by other investigators (Cabrera and Nwaubani (1992)) that single parameters like the average pore diameter (APD) in conjunction with total porosity are good indicators of pore structure characteristics. Therefore in this paper pore size distribution is represented by the APD.

Figure 2. Relation between total porosity and curing time

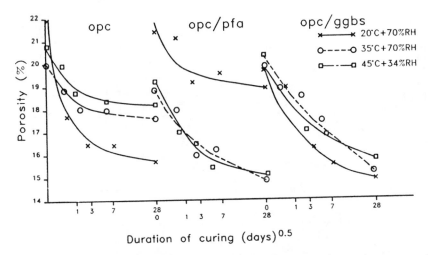

Figure 2. illustrates the influence of duration of curing on total porosity measured at 28 days. A reduction in total porosity with increased duration of curing is clear. Compared to the permeability trends however, the relative reduction in porosity with increased curing periods is significantly less. For example the largest reduction of porosity between conditions of no curing and 28 days continuous curing for the opc mix stored in the $20^{0}C$ 70% RH environment was 6% which represents a change of only 28%.

For those specimens subjected to no curing the highest porosity values were obtained in the $20^{0}C$ and 70% RH environment. This is certainly true for the pfa and opc mixes but in the case of the slag mix the results obtained in all three environments are similar.

Curing periods of one to 7 days in hot environments gave higher porosity. When comparing the relative behaviour of the three different cements it is difficult to detect any clear trend and many of the differences shown in the figure are not statistically significant. What is clear though is that for all curing periods in the two hotter environments the porosities of the samples made from both blended cements are similar and lower than those made from opc.

The influence of duration of curing on the average pore diameter (APD) measured at 28 days is shown in Fig.3. In many respects the results show similarity with the porosity results particularly for the two blended cements containing pfa and ggbs. A reduction in APD indicates that the pore structure is becoming finer and in all cases an increase in curing time leads to an overall finer pore structure. The relative change in APD with increased duration of curing is greater than that for porosity but less than for permeability. Comparing the results for specimens that were not cured with those continuously cured for 28 days the percentage reduction in APD ranges from 14% to 82%. As with previous tests the most sensitive mixes are those made with the opc/ggbs blend which showed reductions of between 55% and 82%.

Figure 3. Relation between APD and curing time

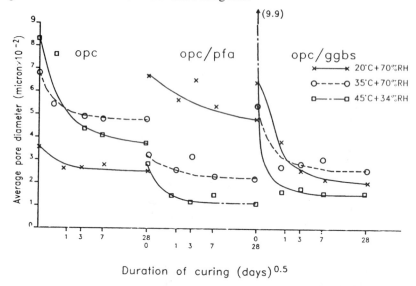

Duration of curing (days)$^{0.5}$

For conditions of no curing and with the exception of the pfa mixes, the most aggressive environment was, as might be expected, that of 45°C and 30% RH. In the case of pfa the specimens kept at 20°C and 70% RH had the coarsest pore structure regardless of the period of moist curing. Certainly for conditions of no curing one would have expected the reverse to be true but this trend (for the pfa mixes) is consistent with that of porosity shown in the previous figure.

Regardless of curing time the specimens made from either the pfa or the ggbs cements and stored in either of the two hot environments were shown to have a finer pore structure than those made with opc. In addition, when compared with that of 20°C and 70% RH, storage in the hotter environments improved the pore structure of the pfa cements. The reverse was true for the Portland cements whereas for the slag cements the influence of storage environment was less significant and less obvious. This trend is consistent in many ways with that of porosity discussed previously.

4 Discussion

In discussing the results and particularly when making comparisons between different mixes it must be remembered that the water/cementitious ratio was kept constant at 0.45. Whilst this helps to simplify the analysis of results it does not reflect what would happen in practise with concrete. It is usual to make concrete to a constant workability rather than constant water/cement ratio in which case on would expect to see a significant reduction in the water content of the pfa mixes (ie up to 15%) and perhaps a marginal reduction for the slag mixes (no more than 5%). If the tests had been made at a constant workability, the pfa mixes would have shown significantly improved properties compared to those reported here whilst the slag mixes would have shown a marginal improvement. The following discussion is however based on the data as presented.

It has been reported by several investigators in the past Goto and Roy (1981), Roy and Parker (1983), Kumar et al (1987) that an increase in temperature has an adverse effect on the pore structure and permeability of mortars or concretes made with opc. This has been attributed to the fact that the rapid initial hydration appears to form products of a poorer physical structure which are probably more porous and therefore a large proportion of them remain unfilled. On the other hand it has been shown that cements made from blends of opc/pfa or opc/ggbs do not appear to be adversely affected by an increase in temperature, Kumar et al (1987), Bakker (1983) and March (1984). Bakker (1983) suggested that the reason for this is that products of the hydration reactions of the pfa and ggbs are precipitated within the pore structure of the hydrated Portland cement leading to a densification of the overall hydrate structure. Any increase in temperature leads also to an increase in the rate of reaction of the pfa or ggbs hydration products. This has the effect of lowering the permeability, reducing the porosity and increasing the fineness of the pores. With regard to the work reported here the influence of temperature alone may be assessed by observing the results of the continuously cured specimens. These are summarised in Table 5. For opc mortars curing in either of the two hot environments has an adverse effect on all three properties measured and the reverse is true for the opc/pfa mortars. The trend for the opc/ggbs mortars is less clear, the permeability and APD measurements show that, compared to 20°C, storage at a temperature of 45°C has a beneficial effect yet the opposite is so at 35°C. Porosity on the other hand is little affected by temperature. No explanation can be given for this reversal of trend between 45°C and 35°C storage but it is interesting to note that for the opc mortars specimens cured at 35°C for more than 3 days the APD and permeability were the highest. See Figures 1, 2 and 3.

There is little doubt that of the mixes tested the one that was most susceptible to duration of curing was that containing ggbs. This is only to be expected in view of the fact that 60% of the cement had been replaced by slag compared with only 30% replacement in the case of pfa. Curing for only one day produced significant improvements in the properties of the slag mixes and after three days curing their properties (as measured by APD and permeability) were similar to or better than those of specimens made from opc.

With regard to critical curing periods the results indicate that the most critical time is about three days regardless of cement type or curing environment. After this period there is little change in either permeability or average pore diameter although the porosity of the pfa and ggbs mixes is still reducing at this stage; certainly any changes that were observed after seven days curing were only minimal. These findings agree to some extent with the recommendations given in the British Standard BS8110 (1985). For Portland cement concretes under "poor ambient conditions" with temperatures in excess of 10°C the minimum curing period recommended is four days. When using cements containing ggbs or pfa it is suggested that the curing period should be extended to a minimum of seven days. On the basis of the results presented by the authors there is no evidence to suggest that the pfa or slag mixes need any more curing than those made with opc. The recommendations given by FIP

Table 5 Properties

Property	Cement type	20°C + 70% RH	Environment 35°C + 70% RH	45°C + 30% RH
Coeff. Permeability (cm/s x 10⁻⁸)	opc	0.32	0.62	0.45
	opc/pfa	0.50	0.48	0.18
	opc/ggbs	0.35	0.75	0.30
APD (micron)	opc	0.0254	0.0584	0.0368
	opc/pfa	0.0474	0.0224	0.0117
	opc/ggbs	0.0205	0.0248	0.0177
Porosity (%)	opc	15.7	17.8	18.4
	opc/pfa	18.7	15.0	15.0
	opc/ggbs	14.7	14.7	15.0

(1986) in their guide to construction in hot weather are that concrete should be cured for a minimum of seven days, no reference is made to the influence of cement type or curing temperature.

It was anticipated at the outset of this project that the environment which would prove to be the most adverse would be that of 45°C and 30% RH. The results obtained by the authors do not confirm this view except that is, for conditions of no curing.

The permeability results consistently show higher values for all three mixes cured in the 35°C, 70% RH environment. This trend is confirmed by the APD results of those opc and opc/ggbs mixes subjected to two or more days of curing. However the APD results of the opc/pfa mixes clearly show the 20°C, 70% RH environment to be the least favourable and the 45°C, 30% RH to be the more favourable.

Although there is no relevant fundamental data with which to explain the trends found, the following explanation is offered:

a)   The rate of a chemical reaction (hydration in this case) is approximately doubled for every 10°C increase in temperature (Cabrera and Nwaubani (1992)), so that even with only one day of curing there is far more volume of reaction products produced at 45°C than at 35°C or 20°C.

b)   Although water evaporates at faster rate at high temperature, the rate-time relation is an exponential decay function which has zero loss at a depth of approximately 70-100 mm. (Gowripalan et al (1989)).

Due to both these facts there is a strong probability that there was sufficient free water for the hydration reaction to continue even when curing was for only one day. This will explain the trends for both pfa and ggbs mixes. For the opc mix the acceleration of hydration creates very rapidly a hydrated shell which surrounds the unhydrated cement grains, this results paradoxically in a reduction of the total hydration, thus at high temperature and a curing time of more than one day the opc mixes gave the poorest results.

5 Relation between permeability and pore structure

The numerical parameters to quantify the durability of concretes and mortars presented in this study have been used to explore relations between them. Although it is accepted that the pore structure of concrete is related to its permeability there is few data to quantify this relationship. The results presented here show that when quantifying pore structure by APD and total porosity a statistical valid function exist between the oxygen coefficient of permeability and pore structure for all the mixes and the three curing environments. This relation is of the following form:

$$\log_{10} OP = 0.5 \; \frac{1}{d^2 + p} + 5.3 \qquad (2)$$

where:

OP = coefficient of oxygen permeability (cm/s) x 10^{-8}
d = average pore diameter (micron)
p = porosity as a fraction

$\dfrac{1}{d^2 + p}$ = pore structure characterising parameter

The coefficient of correlation p^2 is 0.6.

Figure 4 shows a graphical representation of equation (2).

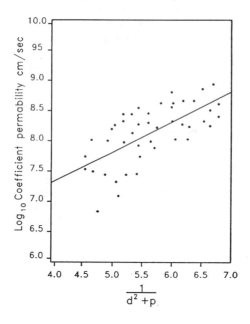

6 Conclusions

From the results presented in this paper, the following conclusions are presented:

1. The permeability and pore structure characteristics as described by total porosity and average pore diameter of opc, pfa and ggbs mortars is detrimentally affected by total lack of curing for any of the curing environments.
2. The hot dry environment (45°C and 30% RH) is detrimental to the durability related properties of the opc mix but favourable to the improvement of properties of the opc/pfa mortar mix.
3. The results presented show that curing for 3 days is an adequate period to obtain low values of porosity, average pore diameter and permeability in any of the curing environments used in this study. Curing periods of more than 3 days cause only a very small change in the properties measured.
4. A statistically valid numerical relation which describes the permeability - pore structure function for all mixes cured in the three environments is:

$$\log_{10} OP = 0.5 \ \frac{1}{d^2 + p} + 5.3$$

5. The opc/ggbs mix proved to be the more sensitive to lack of curing.

6. In the light of these results it appears appropriate to recommend the use of pfa and ggbs in hot climates for the production of concretes of adequate durability.

7 References

Bakker, R.F.M. (1983) Permeability of blended cement concretes. **First Int Conf on the Use of Fly Ash, Silica Fume, Slag and other Mineral By-products in Concrete. ACI SP 79,** 589-605.

British Standard Institution (1983) Coarse and fine aggregates from natural sources for concrete. **BS882:Part 2:1983,** BSI, London.

British Standard Institution (1983) Structural use of concrete. **BS8110:Part 1:1985,** BSI, London.

Cabrera, J.G. (1985) The use of pulverised fuel ash to produce durable concrete. **How to make today's concrete durable for tomorrow. ICE, UK, Thomas Telford, London.** 29-57.

Cabrera, J.G. (1985) The porosity of concrete. **Con Soc Res Sem.,** Leeds.

Cabrera, J.G. and Lynsdale, C.J. (1988) A new gas permeameter for measuring the permeability of mortar and concrete. **Magazine of Concrete Research,** 40, 144, 177-182.

Cabrera, J.G., Gowripalan, N. and Wainwright, P.J. (1989) Curing concrete in hot environments - Assessment of the efficiency of curing membranes by measuring permeability and pore structure. **Third Int. Conf. Deterioration and Repair of Reinforced Concrete in the Arabian Gulf,** 1, 529-542.

Cabrera, J.G. and Nwaubani, S.O. (1992). Strength and chloride permeability of pozzolanic mortars and concretes. Offered to **Magazine of Concrete Research.**

Concrete Society (1991) The Use of GGBS and PFA in Concrete. **Technical Report No. 40 - Concrete Society, Slough, England.**

Goto, S. and Roy, D. (1981) The effect of w/c ratio and curing temperature on the permeability of hardened cement pastes. **Cem Conc Res.,** 11, 575-579.

FIP (1986) FIP Guide to good practice: Concrete Construction in hot weather. **Thomas Telford,** London.

Gowripalan, N., Cabrera, J.G., Cusens, A.R. and Wainwright, P.J. (1990) Effect of curing on durability. **Concrete International,** 47-54.

Kumar, A., Roy, D.M. and Higgins, D.D. (1987) Diffusion through concrete. **Concrete Magazine,** 1, 31-32.

Marsh, B.K. (1984) Relationship between engineering properties and microstructural characteristics of hardened cement paste containing pulverised fuel ash as a partial replacement. **Unpublished PhD Dissertation, Hatfield Polytechnic.**

Reeves, C.M. (1985) The use of ground granulated blast furnace slag to produce durable concrete. **How to make today's concrete durable for tomorrow. ICE, UK, Thomas Telford, London,** 59-75.

Roy, D.M. and Parker, K.M. (1983) Microstructure and properties of granulated slag-portland cement blends at normal and elevated temperatures. **First Int Conf on the Use of Fly Ash, Silica Fume, Slag and other Mineral By-products in Concrete. ACI SP 79,** 397-414.

Senbetta, E. and Scholer, C.F. (1984) A new approach for testing concrete curing efficiency. **ACI Journal,** 81, 1, 82-86.

Wainwright, P.J., Cabrera, J.G. and Alamri, A. (1985) Durability aspects of cement mortars related to mix proportions and curing conditions. **First Int. Conf. Deterioration and Repair of Reinforced Concrete in the Arabian Gulf.** 1, 453-465.

Wainwright, P.J., Cabrera, J.G. and Gowripalan, N. (1990) Assessment of the efficiency of chemical membranes to cure concrete. **Int. Conf. Protection of Concrete. - Dundee, Scotland.** 907-920.

12 PERFORMANCE OF SLAG CONCRETE IN HOT CLIMATES

S. A. AUSTIN and P. J. ROBINS
Department of Civil Engineering, Loughborough University, UK

Abstract
GGBS concrete (with 30%, 50% and 70% cement replacement) and an OPC control concrete, all designed for equal workability and 28 day water-cured strength, were found to have a wide range of strength and permeability when cured under one of four regimes in a simulated arid climate. The effect of curing method and cement replacement level on compressive strength, ultrasonic pulse velocity, water absorption and permeability are reported. The tests showed that a GGBS concrete can have a superior performance to an equivalent all OPC concrete in a hot climate, provided that proper curing is provided. However, without proper curing, the GGBS concrete suffered more than the OPC, resulting in a highly permeable concrete whose strength reduces with increasing replacement level. Of the replacement levels studied the 50% mix produced the best performance.
Keywords: Cement replacement, GGBS, Hot climate, Curing methods, Strength, Permeability, Water absorption.

1 Introduction

Problems are frequently encountered in producing good quality concrete in hot climates. In particular, inadequate curing can result in early age cracking or porous and permeable concrete, which in turn produces structures which are then prone to reinforcement corrosion and other processes of degradation. In short, producing strong and durable concrete in a hot climate can be considerably more difficult than in a temperate one. This has led to an interest in suitable curing methods and alternative cementitious materials to improve concrete quality in hot countries.

At Loughborough we have been investigating the use of cement replacements (ground granulated blastfurnace slag, GGBS, and condensed silica fume, CSF) in hot climates concentrating on curing methods, replacement levels and durability related properties as well as strength, Issaad (1988) and Al-Eesa (1990). To simulate concreting in hot conditions, freshly cast specimens are placed in a climatic room for up to 180 days where the temperature and humidity can be varied over a twenty-four hour cycle period to simulate a particular climate (for example Africa or the Middle East).

The work has compared OPC and OPC modified mixes designed to have equal workability and 28 day strength when cured under water at 20°C; this is in contrast to the practice of simple weight for weight replacement which makes direct comparison of performance less meaningful. Duplicate samples have also been conditioned in the temperate climate of the laboratory to identify the relative efficiencies of the replacement materials in the different climates.

Concrete in Hot Climates. Edited by M. J. Walker. © RILEM
Published by E & F N Spon, 2 - 6 Boundary Row, London SE1 8HN. ISBN 0 419 18090 7.

The research on GGBS has concentrated on the effects of a variety of curing methods on the development of strength and permeability, Austin et al. (1992), together with the effects of cement replacement level and specimen type, Robins et al. (1992).

2 Hot climates and GGBS concretes

In order to obtain good concrete the placing of an appropriate mix must be followed by adequate curing in a suitable environment during the early stages of hardening. Curing problems are exaggerated when concreting in hot weather due to both higher concrete temperatures and increased rate of evaporation from the fresh mix. The durability, strength, and other characteristics of concrete in hot climates are thus critically dependent on its treatment from the moment it is compacted and during the first few weeks afterwards. Inadequate curing can negate all the earlier care taken in mix design and concreting operations, and can also lead to serious defects such as plastic shrinkage cracking and excessive drying shrinkage.

In hot arid countries, as typified by North African and the Middle East, the temperature often rises above 40°C in the shade mainly during the months of May, June, July and August. The mean maximum temperature in the summer day time can be 25°C, and thus a variation in the ambient temperature of 20°C within 24 hours is typical and can be as much as 30°C. This when combined with prolonged exposure to direct sunlight can cause several problems, particularly as a result of the increased rate of surface water evaporation. High ambient temperatures can also raise the temperature of the concrete ingredients to such a level, that on mixing, rapid stiffening occurs, preventing satisfactory compaction. Curing practice in hot countries tends to take more account of the climatic conditions because of the higher temperature and greater sunshine involved. For example, in the Middle East a general awareness of the importance of curing in the production of durable concrete is nowadays reported.

The strength and durability of GGBS concrete, and in particular its resistance to penetration by liquids, gases and ions, are both highly susceptible to curing conditions, more so than a comparable grade OPC mix. This susceptibility can be attributed to reduced formation of hydrates at early ages leading to increased loss of moisture which would otherwise be available for hydration to continue. BS 8110 (1985) recommends longer periods of curing for GGBS concretes in hot conditions and where the surface is exposed to sun and wind. For cement-rich OPC concrete, a substantial rise in temperature during the early stages of hardening has adverse effects on the strength and the durability at later ages. In contrast, the strength and permeability of GGBS concrete have been reported, Roy and Idorn (1982) and Bamforth (1980), to be less adversely affected by such conditions and may even benefit from increased temperature provided that good curing is applied. CIRIA (1984) noted that there had been no systematic evaluation of the performance and specific requirements of concrete made with GGBS in the Gulf region and concluded that field trials and laboratory tests were needed; such studies would allow better specification of concrete and concreting operations.

The initial rate of hydration of slag cements is slower than that of Portland cements, but the strength development at later ages is greater. Several researchers [Pratas (1978), Kokubu et al. (1989) and Chern et al. (1989)] have shown that the strength of GGBS concrete with weight for weight cement replacement is substantially less than that of the OPC concrete at lower temperatures, but that under elevated curing temperatures (greater than 20°C) the strength of GGBS concrete continues to increase at later ages (beyond 28 days) whilst the corresponding OPC concrete strength is constant or decreases. A fairer comparison is made when GGBS concretes are designed for equal 28 day strength and workability, when their later age strength can exceed that of the OPC control (Harrison and Spooner (1986)).

Gowripalan et al. (1990) have recently reviewed the effect of the method and duration of curing on porosity, permeability, and water absorption of GGBS concrete made and found that at 70% replacement it had a lower porosity when cured at 35°C than when cured at 21°C; this illustrates the potential advantage of using cement replacements in hot climates.

Various studies, including that of Graf and Grube (1984), have reported that GGBS concrete can have lower permeability than an equivalent OPC concrete with good curing, but a higher permeability when the curing is poor. Reeves (1985) has also been reported that as the GGBS content is increased, the permeability of concrete decreases. This is the result of the pore refinement of the cementitious matrix through the reaction of GGBS with the calcium hydroxide and alkalis released during the hydration of the Portland cement.

3 Laboratory test programme

3.1 Materials, mix proportions and specimen preparation

Ordinary Portland cement supplied by Castle Cement, conforming to the requirements of BS 12 (1989), was used throughout the test programme. The GGBS was supplied by the Blue Circle Group from its Rouse works near Cardiff, and conformed to BS 6699 (1986), with a chemical composition of 41% CaO, 8% MgO, 36% SiO_2 and 12% Al_2O_3. The fineness ranges of the cement and slag (ex works) were 360-380 m^2/kg and 440-460 m^2/kg respectively. The aggregates were river sand and gravel with a maximum size of 20mm.

To provide a sound basis for comparison of performance, the OPC and three OPC/GGBS concretes (30%, 50% and 70% replacement) were designed to have equal workability and 28-day strength when cured under water at 20°C in accordance with BS 1881 (1983). The mix proportions are given in Table 1. The slumps of all mixes were in the range 45 to 65mm.

Table 1. Mix proportions

Mix reference	Cement (kg/m³)	Slag (kg/m³)	Slag (%)	W/(C+S) ratio	Coarse Aggregates (kg/m³) 20mm	10mm	Sand (kg/m³)
OPC	360	0	0	0.54	720	360	750
30% GGBS	250	110	30	0.53	720	360	750
50% GGBS	205	205	50	0.50	720	360	750
70% GGBS	130	310	70	0.48	720	360	750

The 100mm cube, 500 x 100 x 100mm beam and 500 x 500 x 150mm block specimens were cast in two batches for each combination of concrete type, curing method, and climatic condition. After casting and finishing, all specimens were covered in polythene and, with the exception of the cubes to be cured under water in accordance with BS 1881 (1983), transferred immediately to the climatic room for hot conditioning or into the open laboratory for temperate conditioning, where they remained until testing.

3.2 Curing methods

The range of curing methods was selected on the basis of encompassing the majority of methods commonly used on site. The methods investigated were wet burlap, polythene sheet, curing membrane, air curing and water curing (control cubes only). After 24 hours one of the following curing methods was applied:

1. continuous covering in wet burlap for the first seven days, after which specimens were uncovered until testing;

2. wrapping in two layers of polythene sheet for seven days after which it was removed until testing;

3. brushing on a curing membrane (SBD ritecure 90) to all exposed surfaces; and

4. air curing, that is no proper curing procedure was used.

3.3 Hot and temperate conditioning

The climatic room, measuring 5.5 x 3.3 x 2.9m, was programmed to operate between two 12-hour cycles of temperatures and humidities simulating an arid zone. During the heating phase (day) the maximum temperature actually achieved was $43 \pm 2^{\circ}C$ and the minimum relative humidity was in the range of 25 to 30%. During the cooling phase (night) the minimum temperature was in the range of 10 to 13°C whilst the relative humidity attained its maximum in the range of 70 to 75%. These daily variations are at the extreme of what might reasonably be experienced on site. A typical 24 hour cycle of temperature and humidity is shown in Fig. 1.

In addition to conditioning specimens in the simulated hot climate of the environment room, a duplicate set of specimens was conditioned in the temperate climate of the lab where the temperature was 16 - 20°C and the relative humidity in the range 50 to 60%.

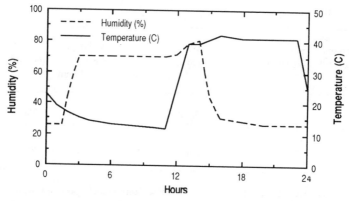

Fig. 1   24 hour temperature and humidity cycle

4 Compressive strength

4.1 Influence of curing

The effect of the four different curing regimes on the strength development (up to 28 days) of the 50% GGBS and OPC mixes conditioned in the hot climate are shown in Fig. 2, together with the polythene cured temperate strengths for comparison. For both these mixes the specimens that were initially cured with wet burlap for 7 days attained the highest strengths at all ages, with air curing producing the lowest strengths. Between 7 and 28 days in the hot climate, the moist cured GGBS concretes gained more strength than the OPC mixes, (an average of 42% compared with 28% for the OPC), whilst the air cured concretes showed only an 8-12% increase.

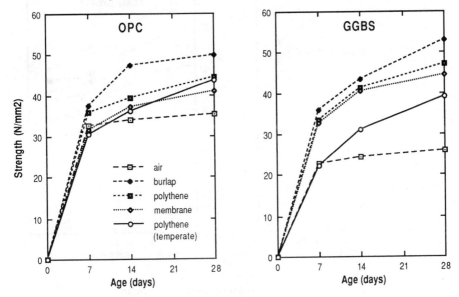

Fig. 2.   Development of cube compressive strength

Comparison of the strength development of the polythene cured OPC concrete in the hot and temperate environments (Fig. 2) shows that hot conditioning increased the 7 and 14 day strengths but had very little effect on the 28 day strengths. In marked contrast, the high curing temperature increased significantly the strength of the GGBS mixes at all ages. This was true for all moist curing methods. Also, the difference between the air and moist cured strengths is much greater with the GGBS concrete, confirming that the use of GGBS can be advantageous when concreting in a hot climate, provided some form of moist curing is provided over the first week.

4.2 GGBS replacement level
Fig. 3 compares the 28 day strengths of all the mixes (0%, 30%, 50% and 70% GGBS) for each curing method in the hot and temperate climates. It should be recalled that all mixes were designed for equal workability and strength when water cured at 20°C. The average 28 day water cured cube strengths were 50.3, 49.8, 49.9 and 49.4 N/mm^2 for the mixes containing 0, 30, 50 and 70% GGBS respectively, and the average densities of all mixes were in the range 2290 to 2380 kg/m^3.

Under temperate conditions the compressive strength decreased with increasing slag content for all curing methods, with the air cured specimens suffering a more marked decrease than the moist cured ones. The decrease in strength was approximately linear. In the hot conditions, however, some of the slag mixes produced higher 28 day strengths with mist curing than the OPC mix, with the 50% replacement level giving the best performance of the replacement levels used, for all three moist curing methods. With no protective curing (i.e. in air), strength reduced with increasing replacement level, in a similar manner to temperate conditions.

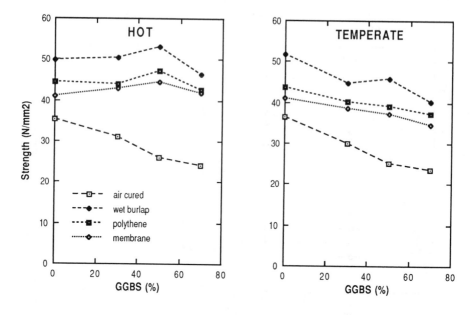

Fig. 3. Effect of slag content on compressive strength

It seems then that GGBS can increase concrete compressive strengths in hot climates provided adequate moist curing is given. However, without proper curing the compressive strength may be significantly below that of an equivalent all OPC mix.

5 Absorption and permeability

5.1 Water absorption

The BS 1881 (1983) 28 day water absorption values (30 minutes immersion) of cores (dried at 105°C) taken from the concrete blocks cured in hot and temperate conditions are compared in Fig. 4 for all four curing regimes. The water absorption was significantly reduced by efficient curing. Setting aside the values for the curing membrane (which will have helped seal the surface), the wet burlap and air curing again proved the most and least effective.

For both hot and temperate conditioning, the slag concretes generally performed better than the OPC when moist cured, but worse when air cured, with again an optimum observed performance at the 50% replacement level. Greatest improvements with increasing slag content were observed for the wet burlap curing, presumably as a result of the extra moisture being made available to the surface of the concrete when curing with the wet burlap over the first 7 days. It is also clear from Fig. 4 that generally, no matter what curing regime was followed, slag concretes performed better under hot conditions, whilst the OPC mix had lower absorptions under temperate conditioning.

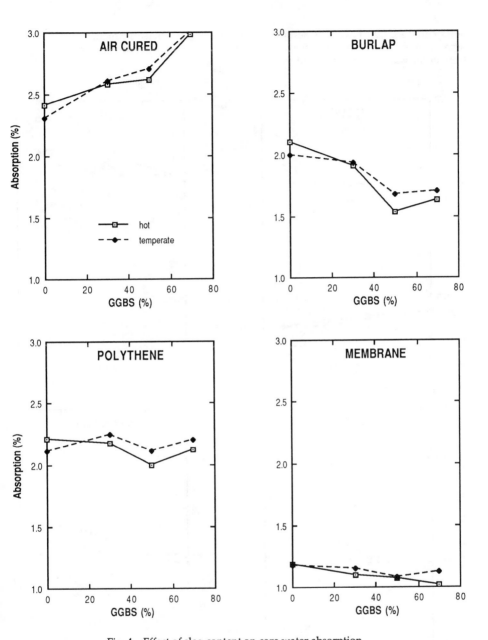

Fig. 4. Effect of slag content on core water absorption

5.2 Air permeability

Air permeability of the concrete mixes was measured using the test method developed by Figg (1973). Air permeability is dependent on the capillary pores in the concrete and since proper curing is an effective way to reduce these, it is not surprising that curing with wet burlap produced the lowest air permeability results (Fig. 5) whilst the highest permeabilities were recorded on the air cured specimens. The trends are similar to those seen for the water absorption results of Fig. 4.

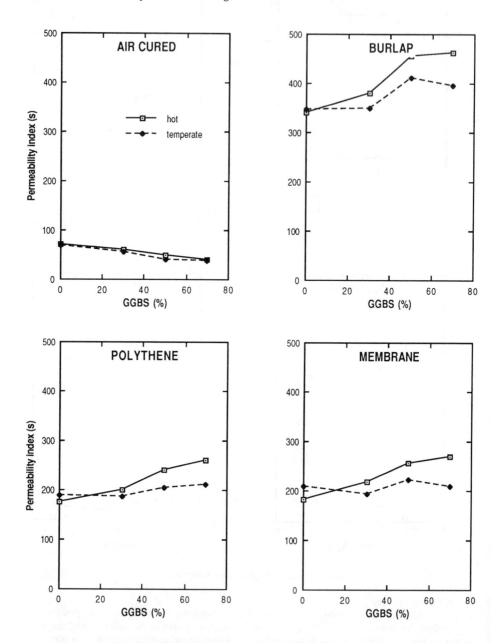

Fig. 5.   Effect of slag content on Figg air permeability index

Under hot conditioning, the GGBS concrete was less permeable than the OPC concrete when it received moist curing. However, when no proper curing was provided, the GGBS concrete suffered more than the OPC concrete and was of poor to moderate quality, based on the Concrete Society (1985) classifications of the index of < 30 = poor, 30-100 = moderate, 100-300 = fair and 300-1000 = good. The higher curing temperature of the hot climate clearly had an influence on the Figg air permeability of the moist cured GGBS concretes, which responded favourably to the increase in curing temperature and exhibited lower permeabilities than those obtained under temperate conditions.

Fig. 5 shows that in the hot climate, provided proper curing was provided, there was a general trend of air permeability decreasing as the slag content increased. In the temperate environment, however, it was only the wet burlap curing that produced a significant reduction in permeability and this only for slag contents of 50% or more. The difference between the hot and temperate air permeabilities of similar moist cured concretes increased with increasing slag content.

6 UPV measurement as a means of monitoring concrete quality in situ

Fig. 6 shows the development of ultrasonic pulse velocity with age for both the OPC and the 50% GGBS concrete mixes cured in the hot climate (together with the polythene cured temperate values for comparison). Comparison of Figs. 6 and 2 reveals that the effect of the different curing regimes on pulse velocity is similar to that on strength: wet burlap and air curing being the most and least effective, respectively. Between 7 and 28 days in the not climate, the moist cured GGBS concretes exhibited a greater increase in pulse velocity than the OPC mixes; this mirrors the trends shown by the compressive strengths of Fig. 2.

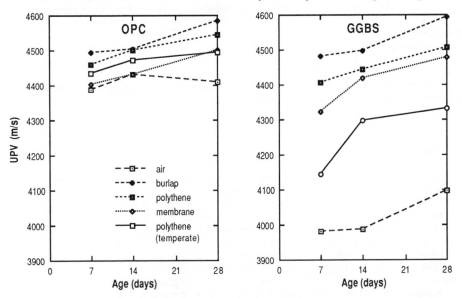

Fig. 6. Development of ultrasonic pulse velocity

Comparison of the pulse velocity development of the polythene cured concretes in hot and temperate environments shows that whereas with the OPC concrete hot curing had only a small effect (and this only at the earlier ages), for the GGBS concretes there is significant difference in values at all ages. Again, this follows the trend exhibited by the strength results.

Though the differences in pulse velocity are modest, the values are what would be expected for the range of concrete strengths observed, Tomsett (1980). The similarity in the trends of the UPV and strength results suggest that UPV measurements could be used on site to monitor curing efficiency quickly and non-destructively.

7 Conclusions

The GGBS concretes and the OPC control concrete were found to have a wide range of strength and permeability when cured under the four regimes investigated in a simulated hot arid climate. Strength, water absorption, air permeability and ultrasonic pulse velocity consistently indicated that wet burlap was the best curing method, that polythene wrapping or curing membrane were also effective (the polythene being slightly better in most cases), whilst taking no precautions produced a very poor quality concrete. Whilst the curing membrane was found to be generally the least effective of the three curing methods studied, it was still satisfactory in producing strong impermeable concrete and may be the most foolproof on site.

In the hot arid conditions some of the moist cured slag mixes produced higher 28 day strengths than the OPC mix; the best of the replacement levels studied was the 50% mix. However, without proper curing, strength reduced with increasing replacement level, in a similar manner to that observed with all curing methods under temperate conditions.

The water absorption and permeability results indicate that in a hot arid climate GGBS is generally less permeable than OPC concrete when it is moist cured, the biggest improvements being achieved with wet burlap curing. However, without moist curing GGBS concrete will suffered more than an equivalent OPC concrete and may be very permeable.

The ability to monitor quality in situ is clearly important and the results suggest that ultrasonic pulse velocity measurement may be suitable for monitoring a structure in the first few weeks following the curing period.

The research has confirmed that GGBS concretes benefit from hot curing. Under these conditions the use of GGBS can produce a superior concrete to an all OPC one in both later age strength and pore structure.

8 References

Al-Eesa, A.S.S. (1990) Silica fume concrete in hot and temperate environments. PhD thesis, Loughborough University of Technology.

Austin, S.A., Robins, P.J. and Issaad, A. (1992) Influence of curing methods on the strength and permeability of GGBFS concrete in a simulated arid climate. Accepted for publication in **Cement and Concrete Composites**, 14.

Bamforth, P.B. (1980) In-situ measurement of the effect of partial Portland cement replacement using either Fly Ash or GGBFS on the performance of mass concrete. **Proc. Institution of Civil Engineers**, 69(2), September 1980, 777-800.

British Standards Institution (1983) Method of normal curing of test specimens (20oC method), **BS 1881 : part 111 : 1983**, British Standards Institution, London.

British Standards Institution (1983) Method for determination of water absorption, **BS 1881 ; part 122 : 1983**, British Standards Institution, London.

British Standards Institution (1985) Structural use of concrete, Part 1. Code of practice for design and construction, **BS 8110 ; Part 1 : 1985**, British Standards Institution, London.

British Standards Institution (1986) Specification for ground granulated blastfurnace slag for use with Portland cement, **BS 6699 : 1986**, British Standards Institution, London.

British Standards Institution (1989) Specification for Portland cements, **BS 12 : 1989**, British Standards Institution, London.

Chern, J.-C. and Chan, Y.-W. (1989) Effect of temperature and humidity conditions on the strength of blast furnace slag cement concrete, in **3rd Int. Conf. on Fly Ash, Silica Fume, Slag and Natural Pozzolans in Concrete**, SP114, American Concrete Institute, Detroit, Vol. 2, pp. 1377-1397.

CIRIA (1984) **Guide to concrete construction in the Gulf region.** CIRIA special publication 31.

Concrete Society (1985) **Permeability of concrete and its control.** The Concrete Society, London, 68.

Figg, J.W. (1985) Methods of measuring the air and water permeability of concrete. **Magazine of Concrete Research**, 25 (85), December 1973, 213-219.

Graf, H. and Grube, H. (1984) The influence of curing on the gas permeability of concrete with different compositions, in **Procs. RILEM Seminar on Durability of Concrete Structures under Normal Outdoor Exposure**, Hanover University, pp. 80-87.

Gowripalan, N. et al. (1990) Effect of curing on durability. **Concrete International**, 12 (2), February 1990, 47-54.

Harrison, T.A. and Spooner, C.C. (1986) **The properties and use of concretes made with composite cements.** C & CA Interim Technical Note 10, November 1986.

Issaad, A. (1988) **OPC and modified GGBFS concretes cured in hot climate.** MPhil thesis, Loughborough University of Technology.

Kokubu, K., Takahashi, S. and Anzai, H. (1989) Effect of curing temperature on the hydration and adiabatic temperature characteristics of portland cement - blast furnace slag concrete, in **3rd Int. Conf. on Fly Ash, Silica Fume, Slag and Natural Pozzolans in Concrete**, SP114, American Concrete Institute, Detroit, Vol.2, pp. 1361-1376.

Pratas, J.D. (1978) **Early age strength development of slag cement concretes.** MSc dissertation, University of Leeds.

Reeves, C.M. (1985) The use of GGBFS to produce durable concrete, in **Improvement of Concrete Durability**, Thomas Telford, London, pp. 59-76.

Robins, P.J., Austin, S.A. and Issaad, A. (1992) Suitability of GGBFS as a cement replacement for concrete in hot arid climates. Accepted for publication in **Materials and Structures**, 25.

Roy, D.M. and Idorn, G.M. (1982) Hydration structure and properties of GGBFS cements, mortars and concrete. **ACI Journal Proceedings** 79 (6), November-December 1982, 444-457.

Tomsett, H.N. (1980) The practical use of ultrasonic pulse velocity measurements in the assessment of concrete quality. **Magazine of Concrete Research**, 32, March 1980, 7-16.

13 THERMAL RESTRAINT IN REINFORCED CONCRETE WALLS AT NORMAL AND HOT WEATHER CONDITIONS

S. M. SADROSSADAT-ZADEH
University of Leeds, UK

Abstract
The restraint factors at different positions of reinforced concrete walls with different L/H ratios have been derived by using Finite Element analysis.The effect of creep on restraint factors has been investigated for normal weather condition.Modification to reastraint factors has been proposed for hot weather condition.It has been shown that restraint factors can be reduced, when concreting at high temperatures.Other conditions being equal,the increase of temperature from 16°C to 38°C can reduce the restraint factors by up to 7%. The effect of time delay between wall and foundation casting has also been studied.It was found that reduction in time delay can significantly reduce the value of restraint factors,particularly for walls with small L/H ratio.
Keywords: Restraint Factor,Temperature,Finite Element Modelling, Creep.

1 Introduction

When cement and concrete are mixed,an exothermic chemical reaction is initiated and heat is evolved.This heat is called the heat of hydration. The evolution of heat of hydration increases the temperature of cement ,or of concrete if aggregate is included in the mix,at a rate which depends on the thermal properties of the concrete and the rate which heat dissipated. Once the rate of heat loss exceeds the rate of heat generation,the concrete starts to cool and to contract.If restrained, as temperature rises,the compressive stresses develop uniformly across the section.These stresses are usually small and partly relieved by creep at early ages.When concrete starts to cool, first any residual compressive stress is recovered,and on further cooling,tensile stress is induced,which again at early ages can significantly be reduced because of the influence of creep.

Thermal cracking occurs when the restrained thermal contraction strain exceeds the tensile strain capacity of the concrete. Determination of the restrained proportion of the thermal movement is essential for economic design,because it is this restrained movement which causes cracking in the concrete.

The present work deals more particularly with the stresses and

Concrete in Hot Climates. Edited by M. J. Walker. © RILEM
Published by E & F N Spon, 2 - 6 Boundary Row, London SE1 8HN. ISBN 0 419 18090 7.

displacements caused by thermal movement in concrete walls connected to concrete foundations cast before the wall. The restraint factors have been derived from the strain and stress distribution at different positions of the wall,for a range of 3 days to 3 months time delay between wall and base casting.Since creep modifies the restraint factors, its effect on restraint factors in normal and hot weather conditions has been investigated.

2 Mechanism of thermal movement at early ages

The build-up of tensile stresses during the cooling period from peak temperature to ambient,can be simply modelled by assuming that the concrete sets at its peak temperature and to ignore the residual compressive stresses which were induced during the temperature rise period.This assumption slightly overestimates the tensile stresses and is therefore a safe basis for design. As a general rule, when the induced thermal tensile strain exceeds the tensile strain capacity, then cracking occurs.

The restraint to thermal movement is the product of the coefficient of thermal expansion of the concrete, the temperature fall from the peak to the ambient and a restraint factor.

2.1 Temperature rise in concrete

As cement hydrates, it generates heat.As the temperature for hydration is raised, the rate of evolution of heat is increases and further raises the concrete temperature.For the usual range of Portland cements Bogue (1955)observed that 50% of the total heat is liberated between 1 and 3 days after mixing with water, 75% in 7 days and 85 to 90% of total heat in 6 months. A record of maximum temperature reached in various structures,due to this heat evolution has been compiled by I. E. Houk(1931),in which he reports a maximum temperature rise of 100°F (38°C).The temperature rise is a function of the balance between the heat generated during the hydration reaction and the heat loss from the concrete,and so depends on the following factors; Quantity and type of cement,Structure size and shape,Mix proportions,Ambient and placing temperature of concrete and Type of formwork and insulation.

The temperature rise data for different section thicknesses and different types of formwork are presented by Harrison(1981).The recorded placing temperature at number of sites in UK shows that on average, the placing temperature in 5°C above the mean daily temperature, while in hot weather with long haul distances, the placing temperature can be over 10°C above the mean daily tempearture.

2.2 Degree of restraint in walls in the uncracked stage

By definition, the stress at any point in an uncracked concrete member is proportional to the strain in the concrete..U.S.Bureau of reclamation (1965) gives guidance on restraint factors for mass concrete blocks for different L/H ratio cast onto mature concrte foundation as presented in fig. 1. ACI Committee 207 (1973) suggests to use multipliers to take into account the relative stiffness of concrete and its restraining foundation. The stress distribution in a

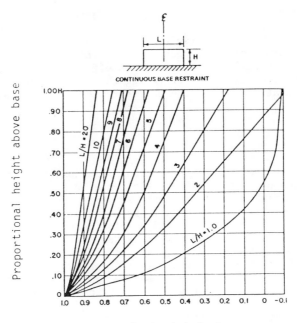

CONTINUOUS BASE RESTRAINT

Proportional height above base

Restraint factor

Fig. 1 Degree of Restraint at Centreline

wall for various L/H ratios,has been analysed by Schleech(1962) .He
also investigated the effect of the development or non–development of
curvature of the wall.Stoffers(1978) suggested a series of curvature
formulae for the restraint factor and showed that in many normal
sections,the restraint factor at base junction can be taken as 0.9,
and at the top of the wall, 0.1.

BS8110: part 2 gives some typical values of restraint recorded for
a range of pour configurations and states that even if a wall is cast
onto a nominally rigid foundation, the restraint is unlikely to exceed
a value of 0.7.

As the stress at any point in an uncracked wall is proportional to
the strain at the same point, determination of degree of restraint needs
either stress distribution or relevant displacements to be established
at that point. Restraint factor for a concrete element may be defined
as:

$$R = \frac{\text{Free Contraction} - \text{Actual Contraction}}{\text{Free Contraction}}$$

Generally the range of restraint factor is from zero to one. Zero when
the element can contract freely,and one when the element movement is
completely restricted by neighbouring elements.In some circumstances
by the effect of movement of neibouring elements, the element may be
elongated rather than contracted and therefore, the restraint factor
exceeds one, or the element may be contracted more than its free

contraction,which means that the restraint factor is less than zero.
   The degree of restraint depends primarily on the relative dimensions, strength and modulus of elasticity of the concrete and restraining material. Any practice which results in limiting the difference in shortening between wall and foundation is effective in reducing the restraint factor.For example, concreting the two parts of the structure in a single operation or with the shortest possible interval of time between them.

3 Finite Element analysis

Theretical analysis of temperature stresses in concrete includes a tedious solution of partial differential equations of diffusion and compatibility. In the recent years the solutions to the problems that have been extremly difficult to solve by means of analytical approaches, have been obtained by numerical computation through the use of computers. In particular the analysis of temperature changes in concrete walls can be performed by the use of Finite Element method. ABAQUS Finite Element program which includes the facility for linear and non-linear analysis, has been used for determination of stresses and displacements at different positions of the walls in the present work.
   ABAQUS also includes the facility for analysis of the temperature changes in the structures.The data input for this type of analysis are:

   The geometry of the structure
   The type and size of the element
   The weight of the elements per unit volume
   The initial and final temperature
   The modulus of elasticity of elements
   The coefficient of thermal expansion
   The types of boundary conditions of the structure

The model used in this study is shown in fig. 2.

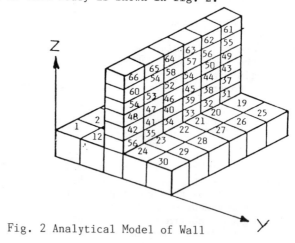

Fig. 2 Analytical Model of Wall

In the present analysis the foundation is divided into 30 and the wall into 36 elements.The length of elements are between 1.5 to 1.25 m for different L/H ratios of walls , following boundary conditions for base have been used ;

1. Base is fixed on the sub-grade (to simulate the case where base is cast on a rock sub-grade
2. Base is free to move horizontally (when base is cast on a gravel soil...)
3. Base is free to curve in the wall plane (when sub-grade is soft and flexible eg. clay soil..)

For each of the above cases , three different length to height ratios for wall (L/H=2 ,3.5 and 5)and for each L/H ratio, the 200mm, 350mm and 500mm wall thicknesses have been analysed, which seems to cover the most practical sizes and dimensions of wall cast onto the foundation. In all cases the thickness of foundation is assumed to be 500mm. Additionally in cases two and three, the 2000mm thicknes for foundation has been analysed as an upper limit.

3.1 Basic assumptions and outline of data used in the F.E. analysis
A number of simplifications have been made in the present analysis;

1.Linear elastic analysis has been made in all cases
2. The temperature-time relationship for walls are based on the case of using 18mm thick plywood formwork which is left in position until the peak temperature has passed in accordance with the data reported by Harrison(1981)
3.The same temperature-time relationship has been assumed for wall and foundation
4.The coefficient of thermal expansion is assumed to remain constant at different temperatures and ages of wall and foundation. An average value of $10*10^{-6}$ /$^{\circ}$C has been used in all analysis
5.The conservative assumption that the concrete sets at the peak temperature in a zero stress state has been made in all cases.

3.2 Effect of time delay between foundation and wall casting
For the estimation of the value of restraint factors at different positions of walls due to restraining action of base, 3 days and 3 months time delay between wall and base casting as a range of practical time delay have been investigated. The modulus of elasticity has been calculated in accordance with BS 8110: part 2,for concrete with f_{cu}= 25 N/mm^2, at relevant ages.The analysis have been carried out in five steps for 3 days and 3 months time delays as shown in fig. 3. The typical data used for Finite Element analysis is shown in fig.3.Some typical results are shown in Appendix A.For full results of F.E. analysis refer to Sadrossadat-Zadeh (1987).

4 Effect of stress relaxation on restraint factors

Relaxation is usually defined as the variation of stress with time at a constant strain. Relaxation influences to a large extent the resistant of concrete to cracking due to temperature changes in concrete.

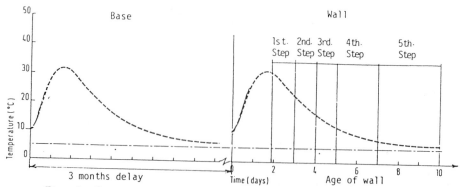

Fig. 3 Step intervals for time delay analysis

Table 1. Data used for F.E. analysis for 3 months time delay

Step	Age of wall (days)		Temperature of wall ($^{\circ}$C)			$E_W *10^{-6}$(kN/mm^2)		
	Start	End	Start	End	ΔT_i	Initial	Final	Ave.
1	2	3	32	23	9	15.5	16.7	16.1
2	3	4	23	17	6	16.7	17.8	17.2
3	4	5	17	12	5	17.8	18.7	18.2
4	5	7	12	8	4	18.7	19.9	19.3
5	7	10	8	6	2	19.9	21.4	20.6
6	10	25	6	4	2	21.4	24.7	23.0

$$\Sigma\Delta T_i = 28$$

Generally the majority of the experimental data on the stress–strain
time relationship for concrete have been obtained in the form of creep
curves,because creep tests are easier to perform than relaxation tests.
The available methods for transforming creep functions into stress
relaxation functions involve rather difficult integral equations but
several approximate procedures have been attempted.

Hansen (1964) considers the case of a specimen subjected at time t_0
to an elastic strain ε_0 ,this strain being maintained constant over the
time interval t_0 to t_1.Initially a stress

$$\sigma_0 = E \varepsilon_0 \tag{1}$$

is required to maintain this strain. If specimen is free to deform
under the applied stress , in time interval t_0 to t_1, it would receive
a creep strain,in addition to the initial strain. Since the specimen
is not free to deform,internal creep will instead relieve a part of the
elastic strain.Thus the elastic strain remaining at any time t_1 is

$$\varepsilon_1 = \varepsilon_0 - \varepsilon_c \tag{2}$$

Where ε_c = creep strain during interval $(t_0 - t_1)$ under a stress
decreasing from σ_0 to σ_1 such that $\sigma_1 = E \varepsilon_1$

Hansen further assumed that creep ε_c, under the decreasing stress can be approximated by the creep that the material would have obtained under the average stress from t_0 to t_1 .i.e.

$$\varepsilon_c \approx \frac{\sigma_0 + \sigma_1}{2} \frac{\varepsilon_c'}{\sigma_0} \tag{3}$$

Where ε_c' = Total creep strain obtained by the test specimen from time of load application $t = t_0$ to $t = t_1$ under constant stress

Substituting eq. 3 in eq. 2 gives

$$\varepsilon_1 = \varepsilon_0 - \frac{\sigma_0 + \sigma_1}{2 \sigma_0} \varepsilon_c' \tag{4}$$

Now since we assume that E remains constant between t_0 to t_1, according to Hooke's law

$$\varepsilon_1 = \frac{\varepsilon_0 \sigma_1}{\sigma_0} \tag{5}$$

Inserting eq. 5 in eq. 4 gives

$$\sigma_1 = \frac{2\varepsilon_0 - \varepsilon_c'}{2\varepsilon_0 + \varepsilon_c'} \sigma_0 \tag{6}$$

Thus if the initial elastic strain (ε_0) and creep (ε_c) under constant stress are known, the initial stress (σ_0) at the end of the interval can be calculated.

The effect of creep on restraint factors has been investigated in a 500 mm thick infill bay with L/H =1/4. Hansen approach has been used for transforming creep functions into stress relaxation functions. The influence of thermal movement and creep on the development of stresses can readily be estimated by a numerical step by step procedure. Therefore, the linear analysis of the infill bay has been divided into five steps, and at the end of each step the effect of relaxation on the stresses has been derived. For each step, the average value of modulus of elasticity has been used and it is assumed to remain constant through each step. The creep coefficient (ϕ_{30}) for appropriate age at 45% relative humidity (recommended value for UK indoor exposure condition) has been calculated from BS8110: part 2 and creep coefficient (ϕ_t) has been calculated from eq. 7 :

$$\phi_t = \phi_{30} \frac{1.04 * t^{0.6}}{(10 + t^{0.6})} \qquad \text{(t, days under load)} \tag{7}$$

The stress relaxations are calculated at wall age of 14 days. The stresses which result from linear analysis (σ_0) at the end of each step has been reduned to (σ_1) because of the effect of stress relaxation.

$$\sigma_1 = \frac{2\varepsilon_0 - \varepsilon_c'}{2\varepsilon_0 + \varepsilon_c'} \sigma_0 \tag{8}$$

Where ε_0 = Elastic strain at the end of each step

ε_c' =Hypothesis creep if the specimen was free to deform during the interval under the given stress (σ_0)

Substituting strain values in eq. 8 with equivalent stresses will give

$$\sigma_1 = \frac{2\dfrac{\sigma_0}{E_0} - \varphi\dfrac{\sigma_0}{E_0}}{2\dfrac{\sigma_0}{E_0} + \varphi\dfrac{\sigma_0}{E_0}} \times \sigma_0 = \frac{2-\varphi}{2+\varphi}\,\sigma_0 \tag{9}$$

According to the principle of superposition, the summation of remaining stresses at the end of each step, has been used as the actual stresses for restraint factor calculation:

$$R = \frac{\displaystyle\sum_{i=1}^{5}\frac{2-\varphi_i}{2+\varphi_i}\sigma_i}{\displaystyle\sum_{i=1}^{5}\frac{\alpha(\Delta T)_i}{L}E_i} = \frac{\displaystyle\sum_{i=1}^{5}\frac{2-\varphi_i}{2+\varphi_i}\frac{(\Delta F_i - \Delta R_i)}{L}E_i}{\displaystyle\sum_{i=1}^{5}\frac{\alpha(\Delta T)_i}{L}E_i} \tag{10}$$

or

$$R = \frac{\displaystyle\sum_{i=1}^{5}\frac{2-\varphi_i}{2+\varphi_i}(\Delta F_i - \Delta R_i)}{\displaystyle\sum_{i=1}^{5}\alpha(\Delta T)_i}$$

Where φ_i =Creep coefficient for step 'i'

ΔF_i =Free contraction of the element during step 'i'
α =Coefficient of thermal expansion
$(\Delta T)_i$ =Temperature change during step 'i'

The results of stress relaxation analysis showed that creep releases the stresses induced by the restraint thermal movement by 45 to 50%. This justifies the reported site values by Hughes (1971) for which a factor of 0.5 on the coefficient of thermal expansion gave a realistic assessment on restraint thermal movement.It makes no different whether this factor is applied to the coefficient of thermal expansion,as suggested by Hughes(1971),or to restraint factors.
The values of restraint factors in table 1 is recommended to be used in UK conditions, when the base is fixed on the sub-grade.
The use of pfa or slag concrete, clearly reduces the temperature rise but they tend to creep less.It has shown by Brooks et al(1991), for equivalent hydration period,the slag concrete creep coefficient is about 30% less than that of OPC concrete.

5 Hot climate concreting and its effect on restraint factors

Hot climate is a general term applicable to countries of relatively high air temperature with low and high relative humidity.From the point of view of the influence of climate on restraint factors,the main factor is creep.As mentioned earlier, during the temperature rise, compressive stresses are introduced in the restrained concrete.These stresses are believed to be small and because of high creep at early

Table 2· Recommended 'R' factors when the base is fixed on the sub-grade

Location	time delay	D=500mm			D=350mm			D=200mm		
		L/H=5	L/H=3.5	L/H=2	L/H=5	L/h=3.5	L/H=2	L/H=5	L/H=3.5	L/H=2
At base	3 days	.48	.47	.43	.48	.46	.45	.48	.47	.46
	3 months	.49	.48	.46	.49	.48	.47	.49	.48	.48
At top	3 days	.38	.21	0	.34	.10	0	.36	.05	0
	3 months	.39	.23	0	.40	.23	0	.40	.19	0

ages,they are mostly relieved.The problem arises during the cooling period when the concrete is more mature and creeps less.

Curing temperature affects the rate of creep significantly.CEB 1970 suggests to replace the age at loading by the corresponding degree of hardening (Maturity), if the concrete hardens at a temperature other than 20 C.Maturity can be worked out by using the following equation:

$$D = \sum \Delta t (T+10) \qquad (11)$$

Where D is the maturity (days C)
Δt represents the number of days during which hardening takes place at T$^\circ$C.
ACI Committee 207(1973) gives the relation between the placing temperature and the age at peak temperature. It also gives the placing temperature against temperature rise in concrete.Since there is no published data on temperature-time relationship of concrete walls in hot climate countries,it is recommended to use ACI Committee 207 graphs to work out the maturity 'D' from equation 11,unless a detailed temperature -time relationship is available. The coefficient K_d for the estimation of creep can then be worked out in accordance with CEB 1970 Code.

The increase of placing temperature which is expected at hot climate concreting increases the creep at early ages.For example,using the above data and Hansen approach implies that the increase of placing temperature from 16°C to 38°C for a wall with V/S = 0.5→1.0, other conditions being equal,decreases the restraint factors by up to 7%.So generally speaking we expect lower values of restraint factors in hot weather concreting.However,the benefit obtained from reduction in restraint factors may well be counteracted by the higher temperature rise at hot weather.

6 Conclusion

The application of 0.5 factor to theoretical elastic restraint factors is an appropriate factor for design purposes in normal temperature conditions .In hot climate conditions,however,a smaller restraint factor can be used in design.Applying the 0.5factor to the calculated elastic restraint factors,shows that the recommended values in BS8110: part 2, for thin wall cast onto massive concrete base, are an

149

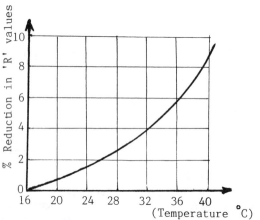

Fig. 4 Effect of temperature on restraint factors

overestimate for restraint factors at the base and underestimate for restraint factors at the top of the wall. According to the results of the present work, the maximum restraint factor suggested by BS8110 at the top of the wall (i.e. R=0.2) can only be used for the sections with L/H ratio not more than 3.5.

The creep modification seems to have already been done on restraint factors suggested in BS8007 (table A.3). These values are very close to the present work results. The restraint factor for top of the wall with L/H=3, however, should be increased from 0.05 to 0.15.

The increase in temperature, which results in higher creep, reduces the restraint factors.As a simple guide fig.4 can be used to modify the recommended restraint factors (table 2) for concrete with placing temperature higher than 16°C.

The use of cement replacement materials for concrete at hot climate, does not guarantee to reduce the risk of early age thermal cracking. In fact the reduction of temperature rise may be counteracted by the decrease in creep.

The effect of time delay between foundation and wall casting on restraint factors can be significant, particularly for walls with small L/H ratio. The effect of time delay is increased towards the top of the wall and is at a minimum at the base of the wall,so having more knowledge about the time delay between foundation and wall casting, is very important in selecting the appropriate values of restraint factors for an economic design.

7 References

ABAQUS Finite Element Package(1984) produced by Hibbit, Karlsson, and Sorensen Inc.
ACI Committee 207(1973) Effect of restraint,volume change and reinforcement on cracking of massive concrete,J.Am.Concrete Inst. 445-470
Bogue,R.H.(1955) Chemistry of Portland Cement.Reinhold,NewYork.

British Standards Institute(1987) Design of concrete structures for retaining aqueous liquids.**BS8007**.

British Standards Institute(1985) The structural use of concrete, **BS8110,part 2**, Code of Practice for Special Circumstances.

Brooks,J.J.,Wainwright,P.J. and Al-Kaisi,A.F.(1991) Compressive and tensile creep of heat cured Ordinary Portland and Slag Cement Concretes.**Magazine of Concrete Research**,43,No.154,1-12.

CEB-FIP(1970) International Recommendations for the disign and construction of concrete structures-Principles and recommendations Comite European du Beton-Federation International de la precontrainte.FIP Sixth Congress. Prague, published by the Cement and Concrete Association,London, 1970.

Hansen,T.C.(1964) Estimating stress relaxation from creep data.**ASTM Materials Research and Standards.** 1,12-14.

Harrison,T.A.(1981) Early-age thermal crack control in concrete. **Construction Industry Research and Information Association**, London, 1981,report 91.

Houk,I.E. (1931) Setting heat and concrete temperatures.**Western Construction News.**

Hughes,B.P.(1971) Control of thermal and shrinkage cracking in restrained reinforced concrete walls.**CIRIA** Technical Note 21.

Sadrossadat-Zadeh,S.M. (1987) Studies of thermal restraint in reinforced concrete walls, MSc dissertation, University of Mancherter,Institute of Science and Technology.

Schleech,W.(1962) Die zwangspannungen in einseitig festgehaltenen wandscheiben(The positive strains in one-way restrained walls). **Beton Und Stahlbetonbau**, 57,64-72.

Stoffers,H.(1978) Cracking due to shrinkage and temperature variation in walls,**Heron**,23,5-68.

U.S.Bureau of Reclamation(1965) Control of cracking in mass concrete structures. Engineering Monograph,No.34,Denver.

8 Appendix A. Typical results of Finite Element analysis

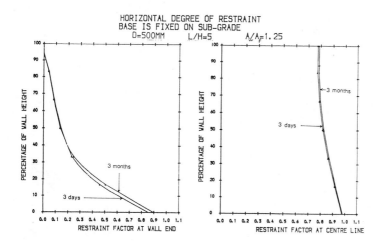

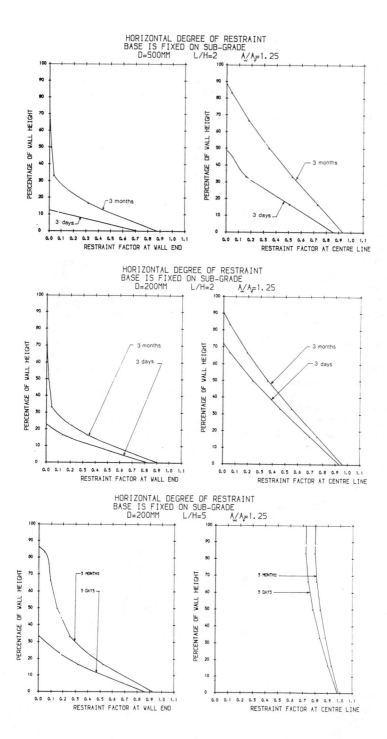

152

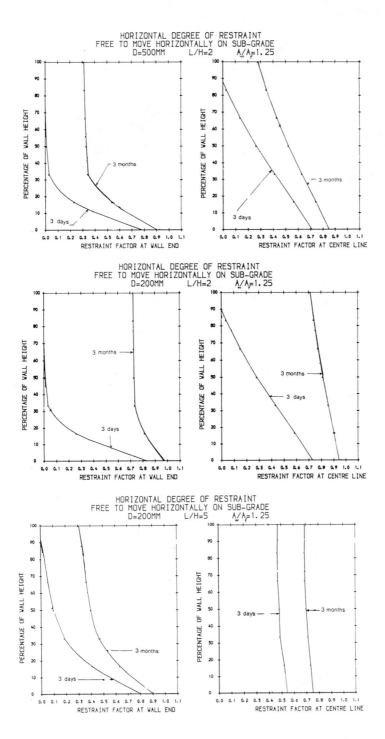

HORIZONTAL DEGREE OF RESTRAINT
FREE TO MOVE HORIZONTALLY ON SUB-GRADE
D=500MM   L/H=2 A_w/A_r=1.25

HORIZONTAL DEGREE OF RESTRAINT
FREE TO MOVE HORIZONTALLY ON SUB-GRADE
D=200MM   L/H=2 A_w/A_r=1.25

HORIZONTAL DEGREE OF RESTRAINT
FREE TO MOVE HORIZONTALLY ON SUB-GRADE
D=200MM   L/H=5 A_w/A_r=1.25

14 LIFE PREDICTION OF REINFORCED CONCRETE STRUCTURES IN HOT AND SALT-LADEN ENVIRONMENTS

S. MORINAGA
Research Institute, Shimizu Corporation, Tokyo, Japan

Abstract
The influences of the ambient temperature, relative humidity and chloride ingress on the corrosion rate of reinforcement in cocnrete were investigated quantitatively. The lives of reinforced concrete structures were estimated based on the life prediction system established by the author. The proper cover thickness and water-cement ratio were proposed which were necessary to secure the required design life under given environmental conditions of temperature, humidity and the distance from the seashore.
Keywords: Reinforced Concrete, Coastal Structure, Durability, Corrosion, Chloride, Hot Country, Life Prediction.

1 Introduction

Durability of reinforced concrete structure is strongly affected by the environmental conditions. In hot countries, the influence of climatic conditions such as the ambient temperature and relative humidity are of fundamental importance. In addition to these, the influence of chloride ingress from the sea is also one of the very important factors. The deterioration is mainly caused by the corrosion of reinforcement, as a lot of papers have indicated many instances of coastal structures which had suffered severe deterioration due to chloride(hereafter chloride is called salt).

Therefore particular measures are necessary in order to make reinforced concrete structures durable in these environments. However the degree of severeness of these environments has not yet fully understood quantitatively. In this paper, the life of reinforced concrete which is determined by the corrosion of reinforcement due to carbonation and due to salt was investigated under various conditions. And the influences of the factors such as temperature, relative humidity, concentration of salt, water-cement ratio, cover thickness of concrete and others, on the life were studied. As a result, proper combinations of cover thickness and water-cement ratio were proposed to secure the required life under hot and salt-laden environments.

Concrete in Hot Climates. Edited by M. J. Walker. © RILEM
Published by E & F N Spon, 2 - 6 Boundary Row, London SE1 8HN. ISBN 0 419 18090 7.

2 Method of life prediction

2.1 The outline of the author's method

The author's method[1] is applied to life prediction of reinforced concrete structures in the case that the life is determined by the corrosion of reinforcement due to carbonation or due to salt. In this method, the life is defined as the time when the corrosion product reaches the critical amount which causes the cover concrete cracking along the reinforcement. In the process to establish this prediction method, the corrosion rates of reinforcements embedded in carbonated concrete or in salt containing concrete under various conditions were investigated. Also the condition for cover concrete to crack due to corrosion of reinforcement was studied.

2.2 The modification of author's method regarding the salt condition

In the case when life is determined by the carbonation, the same method[1] was adopted without modification.

In the case when life is determined by the salt, the modification regarding the salt condition is necessary, because the influence of the salt which is brought into concrete from the external environments is not considered in the method shown in reference[1] In order to take the salt ingress into consideration, the concentration of salt at the surface of concrete and the coefficient of diffusion of salt in concrete must be investigated.

If it is assumed that the salt diffuses into concrete according to the Fick's first law, the equation (1) can be applied.

$$\frac{\delta N}{\delta t} = k \frac{\delta^2 N}{\delta x^2} \tag{1}$$

where
t : time, age of concrete (year)
x : depth from concrete surface (cm)
k : coefficient of diffusion of salt in concrete (cm^2/ year)
N : concentration of salt in concrete at age (t) and depth (x)
  (NaCl kg/ 1 m^3 of concrete)
The solution of equation (1) is expressed by equation (2).

$$\frac{N}{N_0} = 1 - erf (x / 2 (kt)^{1/2}) \tag{2}$$

where
N_0 : concentration of salt at the surface of concrete (x = 0)
  (NaCl kg/ 1 m^3 of concrete)

erf: error function. $erf(u) = (2 / \pi) \int_0^u exp(-t^2) dt$

The concentration of salt at the surface of concrete (N_0) and the coefficient of diffusion of salt in concrete (k) were determined by a procedure shown in section 3. After the value of N_0 and k were determined, the life was calculated by step-by-step method assuming that the concentration of salt in concrete at the age of i-year N_i is equal to the average concentrations of (i) year and (i-1) year, $N_i = (N_{i-1} + N_i)/2$. Except this point, the same method[1] was used.

3 Determination of salt condition

3.1 Surface concentration (N_0) and coefficient of diffusion (k)

The relationships between the depth from the concrete surface and the salt concentration were investigated by literature survey

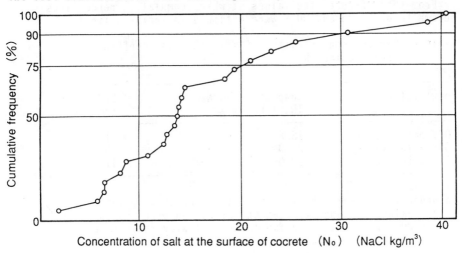

Fig. 1 Cumulative frequency of surface concentration of salt (Zero meter from the sea)

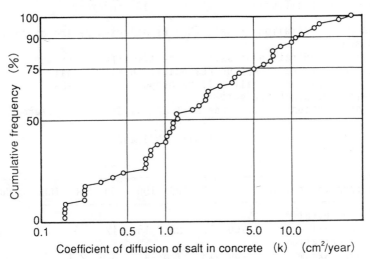

Fig. 2 Cumulative frequency of coefficient of diffusion of salt

157

regarding the forty six specimens which were exposed to the various sea environments around Japan. By adapting these relationships to the equation(2), N_0 and k were determined using the least squares method.

The specimens studied were diversified in age, mix proportion, type of cement, water-cement ratio, specimen size, method of curing and in the locations of exposure site. Some of them were core specimens taken from the existing structures and the details of concrete properties were unknown. Therefore the discussion about the influences of properties of concrete on N_0 and k was abandoned, and N_0 and k were expressed as cumulative percentage diagrams as shown in Fig. 1 and 2. and analyzed statistically.

From these figures, the values at which cumulative percentages became to be 50, 75 and 90 % were obtained as shown in Tab. 1.

Table 1. Surface concentration of salt (N_0) and coefficient of diffusion (k)

Point value	Concentration N_0 (NaCl kg/ m^3)	Diffusion k (cm^2/year)
50% point value	13.8	1.25
75% point value	20.0	5.50
90% point value	29.0	11.5

In order to evaluate the life from a conservative point of view, the higher values of concentration and diffusion were chosen. Referring to the 90 % point value in Tab. 1, the following round values were used in the life prediction.

Surface concentration of salt N_0 = 30 kg/ m^3
Coefficient of diffusion   k = 10 cm^2/year

3.2 Surface concentration of salt (N_0) as affected by the distance from the sea

The surface concentration of salt is greatly affected by the distance from the sea. It may be reasonable to assume that the surface concentration of salt will be proportional to the concentration of air-borne salt.

The concentration of air-borne salt at the distance of zero meter

Table 2. Surface concentration of salt (N_0) and the distance from the sea

Distance from the sea (m)	0	50	100	150	200	500	1000	2000
Ratio of concentration of air-borne salt	100	50	25	15	12	7	5	4
Surface concentration of salt (N_0) kg/ m^3	30	15	7.5	4.5	3.6	2.1	1.5	1.2

from the sea was taken to be 100, and the ratios of concentrations of different points from the sea were obtained referring to the data reported by ISO [2], as shown in Tab. 2. If the surface concentration of concrete at the distance of zero meter from the sea is taken to be $N_0 = 30$ kg/ m^3 as assumed in 3.1, the surface concentrations of concrete at different distances from the sea can be calculated as shown in Tab. 2.

4 Life prediction, factors and levels

Lives were calculated based on the method[1] adding the modification described in section 2. Factors and levels which were used in the life prediction are shown in Tab. 3. The influences of surface concentration were compared at the distances of zero, 50, and 100m.

Table 3.   Factors and levels used in the life prediction

Factors		Levels		
Temperature	(℃)	15,	25,	35
Relative humidity	(%)	50,	70,	90
Water-cement ratio	(%)	40,	50,	60
Cover thickness	(mm)	25,	50,	75
Surface concentration of salt N_0 (kg/ m^3)		30,	15,	7.5
(Distance from the sea (m))		(0)	(50)	(100)

Among the factors which were used in the prediction, the followings were kept constant.
   Coefficient of diffusion : $k = 10$ cm^2/year as mentioned in 3.1
   Type of cement   : Ordinary portland cement
   Finishing materials   : No finish, exposed concrete
   CO_2 gas concentration   : 0.03 %
   Diameter of reinforcement: 13 mm

5 Result and discussion of the life prediction

5.1 Life due to carbonation and life due to salt
The life determined by the corrosion of reinforcement due to carbonation and the life due to salt were calculated and compared. Within the scope of conditions adopted in the calculation, all the lives were determined by the corrosion of reinforcement due to salt, namely, it was confirmed that the attack by salt is far stronger than the attack by carbonation. This shows that the measures against ingress of salt are more important than those against carbonation in hot and salt-laden environments.
   For this reason, the discussion hereafter is referred only to the life determined due to salt.

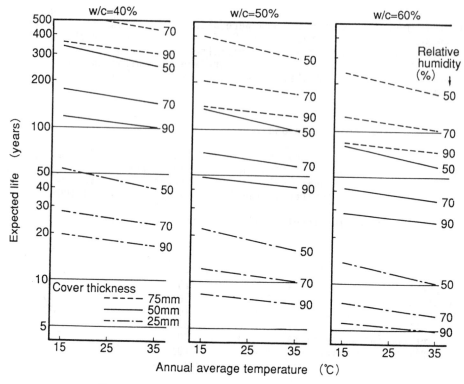

Fig. 3 Influence of temperature (Zero meter from the sea)

5. 2 The influence of temperature on life

Fig. 3 shows the influence of temperature on life. The life decreases as the temperature increases. However the degree of influence of temperature is not so great compared to the other factors mentioned later. Accordingly the temperature is fixed to be 35 °C in the following discussions in the sense to estimate the life from the conservative point of view.

5. 3 The influence of relative humidity

As shown in Fig. 3, the influence of relative humidity is greater than that of temperature. This shows that the rate of corrosion of reinforcement increases rapidly as the humidity increases, and that the more serious measures are necessary in hot and humid environments than in hot and dry ones.

5. 4 The influence of the distance from the sea

The lives at the distances of 0, 50 and 100 m from the sea are compared in Fig. 4. The situation near the sea is very severe. Though the severeness decreases as the distance from the sea increases, considerable severeness still remains even at the 100 m distance from the sea.

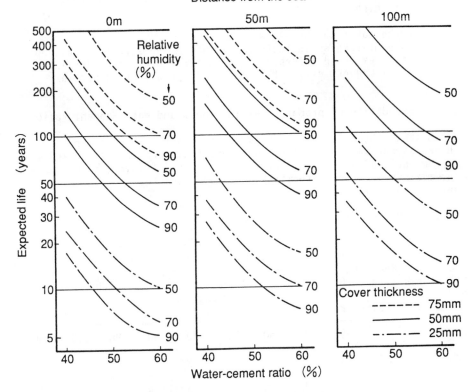

Fig. 4   Influence of distance from the sea

5.5 The influence of water-cement ratio and cover thickness
The water-cement ratio and the cover thickness play the decisive
roles to determine the life as shown in Fig. 3 and 4.

a) When the cover thickness is 25 mm.
If the required life is, say, 50 years, the cover thickness of 25 mm
should not be allowed at the distance of zero meter from the sea
irrespective of the relative humidity. If the combination of 25 mm
cover and water-cement ratio of 60 % is used at this distance, the
life will result in as short as five to ten years.
   Although the life increases as the distance from the sea
increases, the cover thickness of 25 mm is still not enough at the
distance of 100 m from the sea, even if the water-cement ratio is
kept as low as 40 %, when the humidity is high.

b) When the cover thickness is 50 mm.
If the cover thickness is increased to 50 mm, the life is extended
fairly well. At the distance of 100 m from the sea, the life more than
50 years is secured even if the water-cement ratio is 60 % and the
humidity is high. However at the distance of zero meter, the

161

limitation of water-cement ratio is still necessary when the humidity is high.

c) When the cover thickness is 75 mm.
If the cover thickness is increased to 75 mm, the life more than 50 years can be expected irrespective of water-cement ratio, distance from the sea and humidity within the conditions adopted herein.

6 Recommendation of proper cover thickness and water-cement ratio

Fig. 5 shows the relationship between the required life, cover thickness and water-cement ratio at the distances of zero, 50 and 100 m from the sea. Tab. 4 is the data which is read from Fig. 5, and shows the maximum water-cement ratios to secure the required life under several conditions.

Table 4. Maximum water-cement ratio

Cover thickness (mm)	Required life (years)	Distance from the sea (m)								
		0			50			100		
		Relative humidity (%)								
		90	70	50	90	70	50	90	70	50
25	50	N	N	N	N	N	43	N	41	49
	100	N	N	N	N	N	N	N	N	41
50	50	47	52	(60)	55	(60)	(60)	(60)	(60)	(60)
	100	40	43	50	45	49	(60)	49	57	(60)
75	50	(60)	(60)	(60)	(60)	(60)	(60)	(60)	(60)	(60)
	100	53	60	(60)	(60)	(60)	(60)	(60)	(60)	(60)

N : Cover thickness of 25 mm should not be specified.
(60) : Water-cement ratio greater than 60% was obtained.

In Tab. 4, the letter " N " means that the cover thickness of 25 mm is not enough to secure the required life even if the water-cement ratio is 40 %. Therefore under these conditions, the cover thickness must be increased or the water-cement ratio must be decreased to a necessary level less than 40 %.
Also the sign " (60) " in Tab. 4 means that the maximum water-cement ratio greater than 60 % was obtained as a result of life calculation. However it is well known that the higher water-cement ratio may become a cause of various kinds of deterioration. Therefore a certain limit should be specified from an another point of view.

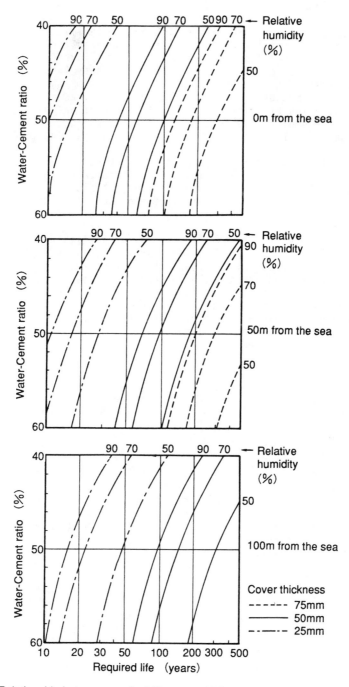

Fig. 5 Relationship between required life, cover thickness and water-cement ratio

163

7 Conclusions

The corrosion of reinforcing steel is one of the main causes to deteriorate the reinforced concrete structures. In the hot and salt-laden environments, the problem is very serious. Investigating the lives of reinforced concrete structures under these conditions, the influense of environments on lives were estimated quantitatively and necessary measures to secure the required lives were proposed.

1) The corrosion of reinfocement due to carbonation and corrosion due to ingress of salt into concrete were investigated. The corrosion due to salt was found to be more serious than that due to carbonation in these environment.

2) As the temperature increases, the life becomes shorter. However the degree of influence of temperature is less compared to the influence of relative humidity. In this sense, the high humidity is more serious than the high temperature, and it is found that structures in hot and humid countries are exposed to more severe conditions than those in hot and dry countries.

3) The influence of salt on life is very great. Although the concentration of air-borne salt decreases rapidly as the distance from the sea increases, it is found that the structures near the sea should be protected against the ingress of the salt.

4) The influences of water-cement ratio and the cover thickness on life are also very great. The proper combinations of water-cement ratio and cover thickness are proposed to secure the required life depending on the environmental conditions and the distance of the structures from the sea.

Acknowledgement
The author acknowledges the devoted efforts of Mr. Tatsumi Ohta, Design division, Nuclear Engineering Department, Shimizu Corporation, who carried out all the work of programming and computation adopted in this work.

8 References

1) S. Morinaga, "Prediction of service lives of reinforced concrete buildings based on the corrosion rate of reinforcing steel", Fifth International Conference on Durability of Building Materials and Components. Nov. 1990, Brighton, UK.
2) Corrosion of atmospheres, ISO/TC, 156, WG4, N 66 E, 1983.

PART THREE

CONSTRUCTION
IN HOT CLIMATES

15 QUALITY ASSURANCE/QUALITY CONTROL FOR CONCRETE IN HOT CLIMATES

N. CILASON
STFA Quality Consultancy, Research, Control and
Auditing Ltd, Istanbul, Turkey

1. What is Quality Assurance and what is Quality Control

Before attempting which word to use "Assurance" or "Control" their definitions must be clear to everybody (to the client, to the contractor and to the consultant).

Quality assurance is defined as " All planned and systematic actions necessary to provide confidence that a service will satisfy given requirements". It is a process of planning or forward thinking. The objective should be, Engineering for quality.

Quality control is " The element of Q.A which is employed to verify such complience. It is the operational techniques and activities. The objective should be inspection for quality.

2. Importance of QA or QC

As both QA and QC will provide confidence that the requirements to produce a concrete that will have enough durability and strength in hot climates, every care should be exercised to implement both of them for the intended life of the structure.

The project management should provide confidence to its own management that the specified quality is achieved.

Any of them or both of them QA/QC is a must for concreting in hot climates. It is not important how you define it, the end product is of vital importance.

3. QA and QC today

Almost every person or organization dealing with concrete have written a paragraph or a page or two pages about quality control and recently, quality assurance.

The importance of QA or QC was tried to be well understood by parties and people dealing with concrete in hot climates. Even some wrote the tasks of the QC department in detail.

Concrete in Hot Climates. Edited by M. J. Walker. © RILEM
Published by E & F N Spon, 2 - 6 Boundary Row, London SE1 8HN. ISBN 0 419 18090 7.

But very little has been said on how to implement these and with what degree of authorization for the quality people.

4   Quality systems

Practical recommendations

4.1 Where quality assurance is requested by the client or decided
    by the contractor :
    The contractor should provide the client with the following
    before the bid or tender :
- Conformation that it operates a quality system conforming to the
  requirements of BS 5750 or ISO 9000 or other international QA
  standard.
- A sample quality assurance manual describing its quality systems.
- Give evidence that its quality management system is documented.

AA. Give typical quality plan.

BB. Give written procedures.

- Details of at least 5 years of how he applied his quality system
  in other contracts in hot climatic regions.
- Details of any quality system audit and approval.
- Evidence that the quality system has been audited and approved by
  any independent body to BS 5750 Part I, ISO 9000 or any other
  international QA standart.

After the tender, the contractor should complete all necessary
procedures for concreting.

The quality control department will be organized within the project
as shown on figure 1.

4.2 Where only quality control is requested by the client or
    decided by the contractor:

The contractor should provide the following before the bid or
tender.

- Evidence that he had a quality control manual and procedures with
  an organizational chart as on figure 2.

- Details of at least 10 years of how he applied his quality system
  to other contracts in hot climatic regions.

After the tender :

- The client or his representative should have a quality control
  engineer having at least the same qualifications or more as of
  the contractor's project QC Manager.

- Project quality manager should have the desired qualifications
  with his team of QC personnel as asked in the contractor's QC
  directive. He should

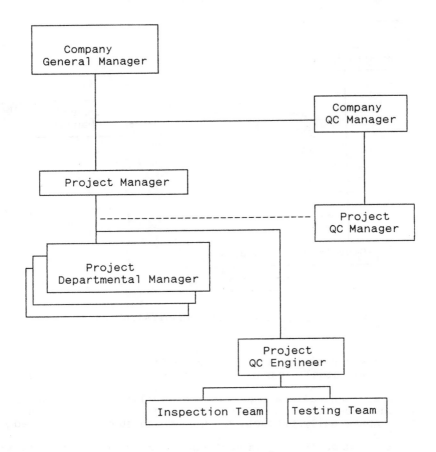

Fig. 1   QUALITY ASSURANCE ORGANIZATION CHART

a) be familiar with drawings and specifications
b) be familiar with construction methods
c) know tests for quality control and be able to teach them
d) be able to cooperate with construction managers
e) be able to inspect units and personnel under his
   responsibility
f) know, how to report and keep files
g) understand that works should comply with the drawings and
   specifications and that he has responsibility in this end
h) exercise his power in responsibility in the best way.

- The project QC Manager should be aware of his duties and
  responsibilities, such as :

  a) to carry out appropriate studies to evaluate possible
     aggregate sources for the project
  b) to carry out field investigations when neccessary, take
     samples and field tests

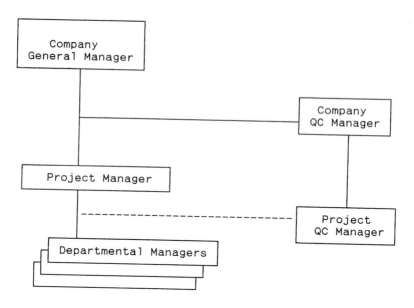

Fig. 2   QUALITY CONTROL ORGANIZATION CHART

c) to collect producer's certificates for materials like cement
   structural steel, aggregate and record delivery and storage of
   them
d) to check and calibrate production facilities for concrete
e) to check performance of transport vehicles for concrete
f) to prepare daily production reports and·submit to related
   offices
g) to see that concreting is continous without unneccesary delays
h) to inspect all the details of production, placement and
   finishing of concrete from the beginning to the end
i) to inspect curing activities and striking times of formwork
   closely
j) to carry out tests to control the quality of production and
   placement of aggregates, cements, steel etc.
k) to have tests on water used for construction
l) to prepare and issue reports for tests
m) to make destructive and non-destructive tests on concrete

- The project QC Manager should be very familiar with testing of
  materials.
- The project QC Manager should have enough power and authority to
  stop concreting activities in the event of persistant lack of
  quality, until proper corrective action is agreed upon.
  He should :

  a) arrange on-site education programs at various levels on
     various subjects for quality control personnel
  b) arrange meeting and remind field engineers and foremen on
     quality related matters

170

c) make studies aiming to raise quality in the works in every respects and prepare reports of findings and suggestions
d) take neccessary steps to stop those activities reporting to the Project Manager
e) inspect and check shuttering, structural steel and others before concreting
f) give consultancy for special concrete works, curing and repairing
g) inspect conditions of storage of structural steel and remind appropriate measures

5. Conclusion

In adition to what has been said until today, an attempt is being made with this report to clarify and to emphasize the importance of the definition of quality assurance and quality control and how it should be applied in project concreting activities.

The difference of the meanings and activities of QA and QC should never be misunderstood or underestimated in the sense that:

- In working with QA, the QC engineer of the project is the manager of the inspection team and the results are verified by testing. That is QC, means inspection.

- In working with QC, the project QC engineer is responsible of searching for proper materials, combining them in the most favorable way and securing the end product from any environmental danger for the specified period and beyond.

Here he is the producer and verifies his activities by testing and inspection.

REFERENCES

1. Concrete in hot countries, Stuvo- FIP
2. Concrete construction in hot weather, FIP guide to good practice
3. The CIRIA guide to concrete construction in the Gulf Region sp. publ. 31
4. Quality assurance in construction, CIRIA sp. publ. 55
5. Quality Assurance and quality control for post-tensioned concrete structures FIP guide to good practice.
6. Some problems of construction aggregates in desert areas investigation, production and quality control. P.G.Fookes, I.E. Higgenbottom, Proc. Instn. Civ.Eng. Part I 1980.
7. Hot weather Concreting Report. ACI Committee 305-1977
8. Recommended practice for concrete inspection. ACI 311-1975
9. STFA Construction Company Quality Control Directive revised 1986
10. STFA (Sezai Turkes- Fevzi Akkaya) Construction Company Quality Assurance manual 1988.
11. BS 5750 Quality Systems

16 PROBLEMS OF UNSUPERVIZED QUALITY CONTROL IN READY-MIXED CONCRETE PLANTS IN SAUDI ARABIA

H. Z. AL-ABIDIEN
Ministry of Public Works and Housing, Riyadh, Saudi Arabia

Abstract
The information provided by some ready-mixed concrete
plants in the Kingdom of Saudi Arabia gives a rosy picture
about the concrete quality and its compressive strength.
The rest of the investigated plants do not even have any
records about their production. However, this research
proves by statistical analysis that the standard deviation
figures provided by various plants are very low in contrast with the international values . Also, the coefficient of variation and characteristic strength showed a very
good quality of the produced concrete. The extensive investigation conducted in this study gave the following
alarming results:
-The majority of values submitted by factories are forged.
-Actual standard deviation of ready-mixed concrete is between 25-90 kg/cm2, while plant provided values were
 5-28 kg/cm^2.
-Concrete characteristic strength in most factories was
 not achieved at least during certain periods of the year.
-Approximately 60% of the concrete produced in the Kingdom
 does not conform to specifications.
-Regarding the existing quality control systems, 33% of
 the examined cases have insufficient and the remaining
 67% have no actual system.
   Since poor quality concrete affects the building safety and causes substantial economic loss, therefore, an effective system is proposed to control quality. The system
is developed from those used in some European countries
and adapted to suit the conditions of the Kingdom. It is
based on self quality control applied by the plant and su-
pervised by independent authorized agencies. Special
attention is given to hot weather concreting.
Keywords :Ready-mixed concrete, Statistical analysis,
Characteristic strength, Hot weather concreting, Internal
quality control, External quality control, Supervisory
agencies.

Concrete in Hot Climates. Edited by M. J. Walker. © RILEM
Published by E & F N Spon, 2 - 6 Boundary Row, London SE1 8HN. ISBN 0 419 18090 7.

173

1 Introduction

The reality about concrete quality and the self quality control systems of the ready mixed concrete plants in the Kingdom of Saudi Arabia was not previously known. Unsystematic external quality control was performed by consultants and some governmental agencies on a limited scale. Hot weather precautions are neither considered nor adopted in the majority of plants. The unavailability of the Saudi specifications for ready mixed concrete added further complications. Due to the occurance of structural damages at early age, from 5-20 years,[1] it was necessary to conduct an extensive study to investigate the actual status of these situations. The following procedures were adopted[2]:
-Completing standard questionnaire
-Site visits, discussions with personnel and inspecting storage methods, maintenance, quality control etc....
-obtaining quality control values.
-Testing samples of concrete constituents.

This study covered 33 ready mixed concrete plants in the Eastern, Central and Western regions of the Kingdom. This represents 40% of the registered plants. Their concrete production reaches about 300,000 m^3 /month.

2 Rosy picture for the statistical evaluation of the plant provided values

In evaluating the values provided by the plants which cooperated by submitting information regarding their self quality control. These values showed marvelous results that can be summed as follows:[3]
-Low standard deviation between 5-28 kg/cm^2
-Coefficient of Variation indicates that 2/3 of these plants produces good to very good concrete. The remaining third of the plants produces acceptable concrete with no plants producing poor concrete.
-All plants achieved a specified characteristic compressive strength (5% fractile).

This rosy picture did not reflect the observed actual status for the inspected plants. However, the extensive evaluation of the questionnaire and the other procedures indicated that only about 12% of the ready mixed concrete plants produces good concrete. While 24% produce acceptable concrete and 64% produce poor to very poor concrete. Also, the study showed that 33% of the plants applied acceptable quality control measures while 15% have incomplete or insufficient quality control. The remaining 52% have very poor or totally non-existent quality control. Therefore, further investigation was essential achieve realistic judgement regarding the quality of concrete production.

174

3 Reviewing the low values of standard deviation

It is important to perceive the accuracy of the standard deviation obtained in the Kingdom. To achieve this, it is beneficial to compare it with some western countries. In some European countries before introducing a comprehensive quality control for concrete, it was found in 1960 that the mean standard deviation for concrete compressive strength is (55 kg/cm^2)[5] . These mean values drop as a quality control system is introduced to (40-45 kg/cm^2)[6,7].

The values obtained for the plants in the kingdom are very close to those for cement alone and not concrete. This means that the effect of aggregates on the standard deviation is excluded. This value is impractical and non-realistic[8,9] , as it is about half less than international values in the developed countries where concrete industry in subjected to rigorous quality control where qualified technicians are available. Hot weather conditions and harsh environment that exist in the Kingdom add further difficulties to the situation.

Various possible reasons leading to the observed low standard deviation in the kingdom were examined.[3] Accordingly, the deliberate manipulation of the results by the plants was perceived to be the main reason.

4 Identifying the actual status of plants in the Kingdom:

To evaluate the actual status of the concrete production the following three methods were adopted:
-Testing concrete samples taken unannounced from the plants' delivery trucks while pouring.
-Evaluating some original log-books for plants which were visited for the second time. These were obtained through considerable difficulties.
-Testing concrete constituents.[2,3,4]
    The results can be summarized as follows:
-The range of standard deviation is very large (27-90 kg/cm) Fig.1 shows the comparison between the actual and the plant prepared values. The difference ranges from 50% to 90%.
-The coefficient of variation indicates that 63% of plants produce poor to very poor concrete while the rest produces acceptable concrete as shown in Fig. 2.
-By examining the concrete characteristic compressive strength (5% fractile) given in Fig.3, it is evident that 20 plants (60% of the plants) produce concrete which does not achieve the characteristic strength during at least some periods of the year. These results are in accordance with those obtained from the questionnaire analysis.

175

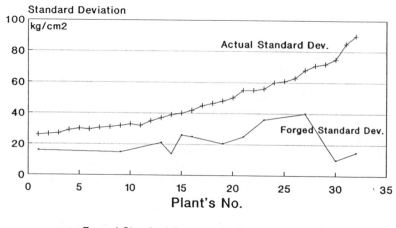

Fig. 1: Forged & Actual Standard Deviation in various Plants

-Hot weather precautions exist in only 16% of the plants and the remainder have insufficient precautions or none at all.
-Since concrete constituents considerably affect both concrete strength and durability, therefore, raw materials were tested for 26 plants (79% of the plants). The test covered cement, water and aggregates.

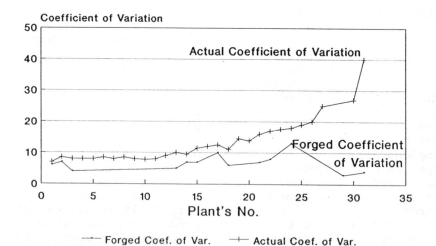

Fig. 2: Forged & Actual Coefficient of Variation in various Plants

176

Fig.4 indicates that 50% of the plants, from which aggregate samples were tested, do not meet the "General Specifications for Building Construction".[10] Generally, it was noticed that in the Eastern region plants use aggregates containing excessive chlorides and sulphates.[11,12,13] On the other hand, in the Western region, plants do not satisfy the soundness requirements. Also, 38% of the plants use water that does not conform to specifications. However, the cement used in various plants meets the requirements of the Saudi Arabian standards organization.[14,15,16]

5 General framework for the proposed system

5.1 Introduction
The system includes both internal self quality control in plants and external quality control from an independent authorized agency. It outlines the general framework for the system and how it differs from those existing in some European countries.[22,23,24,25,26] Plants are classified into two grades. Type (I) produces concrete of strength not exceeding 250 kg/cm^2 and has no special specifications. Type (II) produces all types of concrete with varying specifications. Fig.5 shows the outline for this system and the relationship between the plant, external supervisory agency and contractor (purchaser of concrete)[4]

5.2 Elements of the system
5.2.1 Self quality control
-Responsibilities (responsible manager, engineer, technician, laboratory, conformity)
-Requirements (specifications, repetition and concentration of tests, procedures in case of nonconformity, statistical analysis, sample taking, reporting, hot weather precautions, maintenance, documentation)
-Tests for self quality control for concrete type (I)[4]
-Tests for self quality control for concrete type (II)[4]

5.2.2 External quality control (authorized supervisory agency)
-Responsibilities (supervising agency, validation of internal quality control, tests undertaken by external agency, neutrality)
-General requirements (contract, the engineer and his qualifications, training, prior to production, plant information, site or laboratory inspection, fresh concrete tests, sampling, additional tests, inspection, unannounced inspection, procedure in case of nonconformity, joint report from supervisory agency, documentation).

Plant size	Small	Medium	Large
Production capacity m3/month	<= 5000	>5000 - <=13000	> 13000

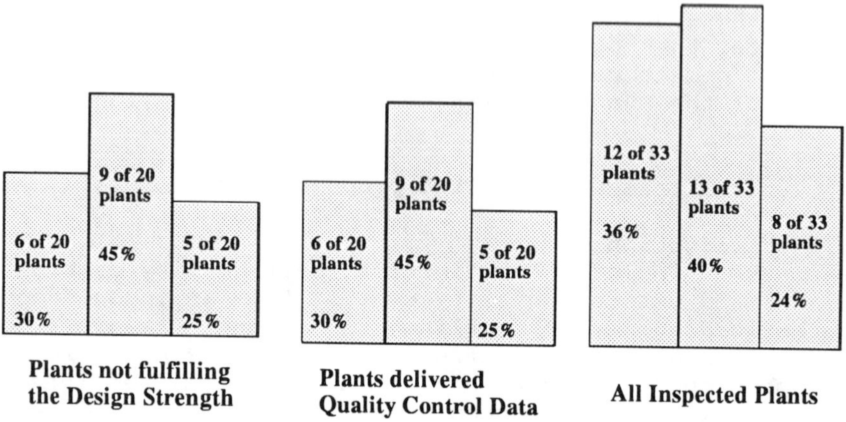

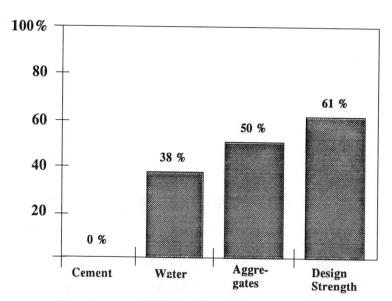

Fig. 3: Distribution Histogram for Plants Sizes,
Fulfillment of Design Strength & Q.C. Data

Fig. 4: Percentage cf Plants not fulfilling
Concrete Related Quality Requirements

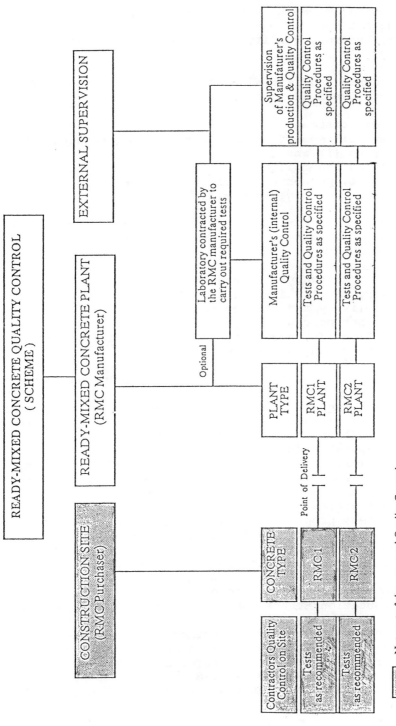

Fig. 5: Out line of the Proposed Quality Control System

-Inspection; Three unannounced visits yearly. At least one during hot summer weather.
-Tests; External quality control tests by supervisory agency[4]

5.3 Suggested additional tests by the contractor for concrete types (I) and (II)[4]

5.4 Specifications

5.4.1 General specifications for plants producing concrete types (I) and (II).

It includes the following:
-Management, personnel, training, office
-Equipments, storage, material handling
-Mix proportions
-Specifications for the ready mixed concrete constituents (Cement, aggregates, water, admixtures)
-Fresh concrete
-Production provisions
-Requests, delivery and receipt (Delivery catalogue, quality bulletin, request components, delivery and receipt report)
-Hot weather precautions (Identification, plant precautions, precautions during batching, mixing and transportation)
-Special conditions for quality control

5.4.2 Special requirements for plants producing concrete type (II)

Additional requirements concerning production, equipments, management and personnel, raw material, quality control and hot weather placement of concrete.

5.4.3 Disputable delivered concrete

Generalities, procedures in case of lost test results, procedures for low concrete strength at origin, procedures in case some properties are not satisfied.

5.4.4 Hot weather precautions

-Aggregates; proper shading, spraying and cooling silos.
-Mixing water; properly isolated or underground tanks, temperature should not exceed 10° C in the hot weather period by applying measures stated in the specifications.[4]
-Production; should not exceed 38° C at pouring time.
-Transportation; trucks should be isolated and painted white
-Additional hot weather precautions for concrete type (II) Special precautions to be taken regarding storage, mixing, transportation of concrete, i.e. the temperatures of mixing water should not exceed 5° C and concrete temperature at pouring time should not exceed 30° C. [4]

6 Adaptation and adjustment of the European system to suit developing countries

6.1 Introduction

The human experience is the basis from which any contribution, in any field, originates and benefits. Nevertheless, no experience should be transferred identically from one to a different country with a different society and distinct circumstances. Every country's environment, circumstances and conditions should be taken into consideration to achieve the ideal adaptation for the experience. Accordingly, the European system of quality control in ready mixed concrete plants, has undergone various adjustments to suit the Kingdom of Saudi Arabia and other similar countries. The system can be developed to suit any developing country by using the same method.

6.2 Bases for developing the proposed system to suit the Kingdom

First: The significant frameworks of the European system has been procured [27,28,29,30] Numerous details and differences related specifically to the Kingdom has been introduced depending on the extensive adopted experience and the results of research and investigation in ready mixed concrete plants.

For example the European system suggests two unannounced visits to the plants per year, the proposed system obligates three such visits per year. The European system allows a supervisory neutral agency to be official or unofficial, the proposed allows only an official supervisory agency.

Second: The proposed system for the kingdom is more comprehensive. This is essential in the preliminary stages as developing countries require more details to serve the capabilities to implement the system and assist those in charge for applying the system.

Third: The classifications of the plants into two levels, which is similar to the European one, was found to be suitable to the Kingdom conditions. However, some modifications were taken into consideration, such as; European system does not allow the use of admixtures in plants type (I), whereas the proposed system necessitates the use of retarding materials in all cases specially during hot weather.

The European system designates minimum cement content related to the aggregate gradation, the proposed system sets one minimum limit for the cement content (350 kg/m^3). See table 1 for more details.

Fourth: The European system relies on each country's spe-

Table 1: Comparison between the European System and the proposed Arabic System for Quality Control on Ready-Mixed Concrete Factories

Subject	European System	Proposed Arabic System
Concrete I: Prescribed Concrete **Self Control**	Admixtures are not allowed	It is necessary to use retarder, specially in the summer
	Minimum cement content according to type of gradation	Min. cement content 350 kg/m3. Agreement with supervisory neutral agency on mix proportions & changes whenever required.
	Reference to various specifications for items	Due to lack of Saudi specifications, therefore detailed specifications were developed for ready-mixed concrete and guide lines for material storage + handling and for maintenance were given
	Limits, periods and requirements are neither specified nor detailed	Fully detailed limits, periods and requirement were specified
	There is no big distinction between concrete I and II Specifying water/cement ratio for both types	Clear separation and each type has its own schedules regarding tests, requirements, limits and specified water/cement ratio for type II (only)
	Limits are not tied to month or week but given generally as continuously, whenever needed , etc ...	Specified by year, month, week or day
	Tests are not required for some concrete constituents like cement and admixtures	Test of constituents is required by neutral agency occasionally. Also the impact of admixtures on concrete has to be reported.
	No tests in relation to temperature are required	Requirements exist regarding mix temperature, mixing water temperature, ...etc
	No partaicular precautions for hot weather	Detailed precautions for hot weahter are given
	No distiction between regions regarding tests	Tests regarding existing materials are regionally different (Eastern, Central and Western Region)
	Number of samples is very limited: concret I 3 cubes for 500 m3 and at least once/month and for concrete II 6 cubes/month.	Bigger number of samples: 9 cubes for 500 m3 and at least once/week, 9 cubes for 500 m3 and at least once/day for concrete of special properties or with strength exceeding 250 kg/cm2
External Neutral Control **(External Supervision)**	At least 2 unscheduled visits per year	At least 3 unscheduled visits per year
	Could be governmental or non-governmental	Governmental only
	The supervisor (inspector) has to check and insure the self control and he has the authority to carry out tests or to request additional supporting tests	The supervisor (inspector) has to verify following specific details: * Cement: — Check factory's delivery ticket — Verify factory's receipt — check proper lables on cement silos — To insure execution of cement tests (for concrete II) — Check proper storage * Aggregates: — Check delivery tickets — Execute visual inspection — Verify the self control test performed by the plant — Inspect moisture content of aggregates. * Same procedure for the other concrete constituents
	No compulsory test from the suprvising agency	Specific compulsory test must be conducted by the exeternal supervising agency
	In case of doubts no details are given	In case of doubtful concrete specifics are given to evaluate concrete compression strength
Specifications and its concedration to the environment	There are no specifications defined for Ready Mixed Concrete	Detailed specifications are available taking into consideration certain circumstances of K. S. A. which became evident through site visits. Questions: - Prevention of fine dust - Protecting water against dust, salt and algae - Personnel training has to be given special attention and should include engineers, technicians and administrators - Archiving test and material documents - Precaution against manipulating the statistical data and calculating reasonable limits and factors - Practical suggestions for both concrete I & II regarding lowering the mixture's temperature during mixing and transportation process

cifications whenever needed. Due to the lack of such specifications in Arab countries, the following steps are taken:
-The preparation of a detailed specification for ready
 mixed concrete in the Kingdom (33 pages).[4]
-The preparation of equipments, materials and storage
 schedule (12 pages).
-The preparation of maintenance instructions (4 pages).
-In case the previous documents are not sufficient, then
 references available in the Kingdom i.e. Saudi Arabian
 Standard Specifications and others[10] are indicated. Also
 international references i.e.ASTM, DIN, BS, ISO..etc are
 given whenever needed.
Fifth: The European system did not deal with specific details as in the proposed system as follows:
-The European system does not differentiate in the tests
 between the two types of concrete, whereas the proposed
 system achieves that in a detailed manner.[4]
-The European system does not specify the tests performance dates or times. In the proposed system day, month
 and year are specifically designated.
-In the European system no tests are required for concrete
 constituents as they are properly controlled by the
 supplier.
The proposed system covers this by additional tests.[4]
-Relating some tests to predominant hot temperatures.
-Hot weather precautions are not included in the European
 system.
Sixth: The European system does not differentiate between
various regions. In the proposed system this has been taken into consideration for various tests according to nature of materials and the prevailing environment.
Seventh: Increasing the number of samples in the proposed
system due to the large discrepancy detected in the concrete industry in the Kingdom.
Eighth: No details are available in the European system
regarding the supervisory agency, whereas the proposed
system points out the specific responsibilities, duties
and tests to be undertaken.[4]
Ninth: Accommodating the Kingdom's circumstances and environment in the specifications:
-Considering hot weather conditions
-Stating the method of treating disputable concrete and
 cases where the strength or specifications requirements
 are not met
-Training of technical personnel from engineers, technicians, administrators etc...[4]

7 Recommendations

-The proposed quality control system and ready mixed concrete specifications should be discussed by all concerned parties involved in construction industry. Any viable modifications should be included and the supervisory agency designated. Accordingly, the system should be approved and applied.

-The factual standard deviation and statistical data obtained from this study should be considered while evaluating plant data. Thorough examination is essential since plant data could be manipulated.

-Particular care should be given to quality and storage of materials used in the concrete production.

-Special care should be given to hot weather concreting by using admixtures to facilitate workability.

-Proper storage and handling of aggregates and water in hot weather should be maintained.

-Issuing simplified manuals and general education for those involved in the concrete industry. Courses and seminars should be conducted for technical personnel.

8 Acknowledgment

The author wishes to thank "King Abdul Aziz City for Science and Technology" for its financial support of this research.
The author also thanks the researchers who contributed in completing this study.

References

/1/ Al-Abidien, Habib M. Z. "Insight on Strucutral Safety and Age of Concrete Buildings in Saudi Arabia"' International Seminar on the Life of Strucutres, British Cement Association, Brighton, England 24-26 April 1989.

/2/ Al-Abidien, Habib M. Z. & Others Quality of Ready-Mixed Concrete in K.S.A. First report Nov. 1989

/3/ Al-Abidien, Habib M. Z. & Others Quality of Ready-Mixed Concrete in K.S.A. Second report May 1990

/4/ Al-Abidien, Habib M. Z. & Others Quality of Ready-Mixed Concrete in K.S.A. Third report April 1991

/5/ Entroy, H.C.: The Variation of Works test cubes. Slough cement and concrete association Research report 10, 1960.

/6/   Metacalf, J.B. The Specification of Concrete strencgth, Part II, Crowthorn
       Road Research Laboatory, Report LR 300

/7/   International Information given by : Federal Institutes for testing Building
       Materials in Munich, Stuttgart, Germany & Graz / Austria

/8/   Wiss, Janny, Elstner Association: Results of Statistical evaluations for
       some ready-mixed concrete producers in U.S.A., 1989

/9/   ACI 214-77 (Reaproved 1983)
       Recommended Practice for Evaluation of Strength Test Results of Concrete

/10/   GSBC General Specification for Building & Cosntruction, Ministry of
       Public Works & Housing, Riyadh, K.S.A., 1st Editon 1982

/11/   Al-Abideen, H. Properties of the Aggregates for Concrete in Saudi Arabia
       "1st Saudi Engineering Conference, Jeddah 1983

/12/   Al-Abideen H. Aggregates in Saudi Arabia: A Survey of their
       properties and suitability for Concrete, Material & Structures, vol 20,1987

/13/   Al-Tayyab, A.J. et al.: Development of building and construction
       materials using available resources in Saudi Arabia,final Report, SANCST AR- 4,1985

/14/   SASO Standard # 142, physical & mechanical tests for
       Portland Cement

/15/   SASO Standard # 143, Normal and fast hardning
       Portland Cement

/16/   SASO Standard # 2047, (draft) Sulfat resistant Cement

/17/   ISO 9000 and EN 29000 " Quality Management and Quality Assurance Standards -
       Guidelines for Selection and Use"

/18/   ISO 9001 and EN 29001 " Quality System - Model for Quality Assurance in Design/
       Development, Production, Installation and Servicing"

/19/   ISO 9002 and EN 29002 " Quality System - Model for Quality Assurance
       in Production and Installation "

/20/   ISO 9003 and EN 29003 " Quality System - Model for Quality Assurance in
       Final Inspection and Test "

/21/   ISO 9004 and EN 29004 " Quality Management and Quality System - Elements "

/22/   DIN 1045 "Beton und Stahlbeton : Bemessung und Ausfuehrung"

/23/   DIN 1084/3 "Ueberwacung (Gueteueberwachung) im Beton- und Stahlbetonbau;
       Teil 3 Transportbeton

/24/   EN 199 (Draft) " Ready-Mixed Concrete Production and Delivery "

/25/   Eurocode 2 " Common Unified Rules for Concrete, Reinforced Concrete and Prestressed
       Concrete Construction / Chapter 6.4 Quality Control "

17 THE SELECTION OF HYDRAULIC CEMENTS TO SATISFY THE REQUIREMENTS FOR CONCRETE CONSTRUCTION IN HOT CLIMATES

P. L. OWENS
Quality Ash Association, Rugby, UK

Abstract
This paper describes for specifiers, manufacturers of cement and producers of concrete, the necessity of increasing the versatility of cements, via a 2 Pack system, where the proportions of the various cement components available can be varied by design at the concrete mixer, to achieve the most durable concrete possible for a hot climate. The paper references appropriate materials and techniques, not only for the appropriation of suitable secondary materials, but for their assessment in combination with Portland cement.
Keywords: Portland Cement, Composite Cements, Secondary Materials, Water Content, Durability, Test Methods, Hot Climates, Concrete Specifications.

1 Introduction

The appropriate selection of a single hydraulic binder (cement) for use as a general purpose concrete construction material for all applications in hot climates is not possible. For this is not just a case of selecting one cement, made to satisfy tests performed under standard laboratory conditions at ambient temperatures of, say 27°C or thereabouts, but very particularly is one of suiting the choice of the cement to its application. Cement specifications are usually ambiguous because of the requirement to satisfy standard tests which are used to judge a cement's conformity to the manufacturing and standard specification, for the purpose of attestation, in preference to assessing the performance of that cement's fitness for purpose.

2 Principles

The first question in the selection of cement must be: Does the concrete specifier have a clear understanding of the concrete practice that operates in the particular hot climate, because of

Concrete in Hot Climates. Edited by M. J. Walker. © RILEM
Published by E & F N Spon, 2 - 6 Boundary Row, London SE1 8HN. ISBN 0 419 18090 7.

the increased rate of reaction with temperature? If not, then the risk is increased of poor performance of the concrete in which any cement will be used.

However, there are two major influences that motivate the specifier. Firstly, the honest desire to do a "good" job and secondly, to capitalise on evidence based on successful practice. This latter aspect is a matter of receiving adequate guidance from various codes of practice, **together with** personal appraisal of the actual experience of **failure**, which tend to be more spectacular in hot climates. For example, the use of low C_3A cements (sulphate resisting Portland cements with less than 3.5% C_3A) in regions where high water tables exist adjacent to a marine environment. Such misapplications of cement have resulted in some very spectacular failures resulting in very expensive remedial measures for reinforced concrete, caused by the inability of the hydrated cement paste to form calcium chloroaluminate from the reaction of C_3A and any chloride present. This is because C_3A in the hydrate has the propensity to intercept and complex chloride, thereby reducing the risk of steel corrosion. Conversely, with a cement containing more than 10% C_3A, then there is increased risk of its conversion to ettringite or internal sulphate attack - called delayed ettringite formation (DEF) as a consequence of high temperatures generated during early age hydration.

The task of selecting the most suitable cement for a particular application is confused by the specification of concrete strength, which is used not only for design purposes, but also for judging compliance with a specified water:cement ratio or cement content. The concept of a concrete specification based on strength can be regarded as one of the most dangerous propositions for the specifier, if a clear understanding of the purpose and application of that concrete is not obtained from the outset. Nevertheless, pragmatism is beginning to recognise that concrete specifications have to specify, not only the cement type and strength class, but also the cement content, as a fixed quantity with the water content limited to a maximum, with the requirement for a minimum workability (slump). This accepts that the manufacture of concrete, in its fresh state, has to be workable enough to be moulded to shape, but where water will not be used to increase the placing rate. This concept, although not new, presents the greatest of all challenges to the specifier of concrete because chemical admixtures have to become admissible.

In essence, and there is a conflict of judgment, it is how to specify and keep the resolution of a complicated process influenced by temperature, time and "man" contained to within a simple and basic text, for concrete production and placement to succeed. By over-simplification and to allocate the mass production of cement to a single specification, has created built-in inflexibility that is impossible to accommodate across the full spectrum of concrete production, which is made at the rawest level, volume batching and banker board, through to the most sophisticated, a fully automatic batching plant with two or more inter-connecting cement silos. Inevitably the problem of versatility has to be resolved and while for the unsophisticated, a "2 Pack" composite cement system may create temporary difficulties, it is the most versatile system available. This is even in inaccessible areas where there are poor facilities, or where the infrastructure is either primitive or even non-existent. Such a proposal may not at first seem appropriate until the economics are fully examined and recognised, because use has to be made of every appropriate resource or secondary material across the "menu" available for any type of fly ash or industrial and natural pozzolana, through to the various types of metal slag.

It is appropriate to note that not only in the European Committee for Standardisation (CEN), but also in ASTM, proposals are currently under consideration by the respective cement committees for the utilisation of every appropriate industrial by-product, by moving away from the general prescriptive and restrictive type of cement specification to those that are performance orientated. It is in this area of cement industry enterprise that both the specifier and the concrete producer has not only to learn, but also to take advantage, for composite cements are being realised as the most appropriate option environmentally, commercially and in application, as they generally provide concrete of greater potential durability, particularly in hot climates.

3 An appreciation of a cement's role in concrete

The proposals to be offered here rely, however, on the traditional production of Portland cement, or its clinker, to be used as the activator or catalyst, for appropriate use with a secondary material; the use of chemical admixtures not only to encourage controlled setting times, but also, and more importantly, to reduce the water content of all concrete to below 150 litres/m^3 without any sacrifice to workability, in fact quite the opposite, the intention is to increase workability.

The principal reason for requiring to lower the water content of all concrete to below 150 litres/m^3 is to ensure aqueous impermeability. This impermeability is not, however, required totally to reduce attack from external agents, but to minimise the risk of leachates from the concrete getting into the aquafiers. For, unless there is commitment to minimal risk of exposure to toxic leachates, environmental legislation will prevent application of appropriate fit for purpose secondary materials, which are being amassed at an unpredicted scale throughout our planet, as its population increases expotentially, creating its demand in turn for industrial products.

4 Performance of fresh and hardened concrete in hot climates

While it was necessary to state in the Principles that a "2 Pack" composite cement system would give increased versatility, it was also necessary to argue control of water content as an alternative to water:cement ratio and its control via strength, so before detailing the requirements and reasons for improving cement performance, the following summary of the effects of hot conditions on concrete and its performance under these conditions will serve as a reminder of what are quite singular but may not be well known features affecting concrete.

Temperatures greater than 25°C cause:

Increases in the water content to sustain workability.

Acceleration of setting and stiffening .

Decreases in workability increase the risks of inadequate compaction and the formation of "cold" joints.

Higher initial placing temperatures which in turn accelerate the rate of hydration which advances both the rate and amount of temperature rise, particularly in larger elements.

As a consequence of higher initial placing temperatures the risk of thermal cracking is increased and damage to the hydrate structure, with increased occurrence of micro-cracking.

That particularly with Portland cement concretes there is:

Risk of lowering the long term strength.

Increased permeability

Risk of internal sulphate attack (delayed ettringite formation - DEF).

Contrary to tradition where the application of concrete with low water:cement ratios, ie. less than 0.45, was thought to be a simple provision, enough to compensate for any deleterious effects by preventing attack from aqueous solutions containing sulphates and chlorides, there is now considerable undisputed evidence of damage to the hydrate structure of Portland cement concrete. This damage, identified by Bamforth (1980) and Richards and Buttler (1981) is caused by the temperature developed during hydration, which compounded by the higher temperatures of the fresh concrete is a matter that now has to be addressed.

Firstly, unless the fresh concrete is conditioned to tolerate temperatures up to 40°C and secondly, designed to generate temperatures less than $10°C/100kg/m^3$ cement content, there is where the least size of an element is greater than 400mm, a serious risk of boiling taking place, particularly where Portland cement contents greater than 400 kg/m^3 are used.

Bamforth (1986) convincingly argued the case for composite cements in hot climates with cogent evidence of the major differences in the strength of concrete affected by temperature (Figure 1). Owens (1985) demonstrated the beneficial effect of higher concrete temperatures with a composite cement, by examining the effects of a pulverised-fuel ash type of fly ash on

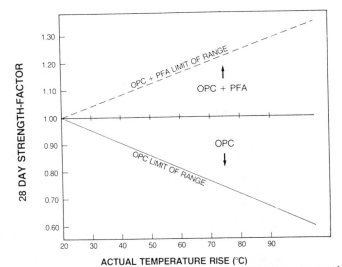

Fig 1 Effect of temperature rise on the 28-day strength of concrete made with Portland cement, with and without pfa. (Bamforth 1986)

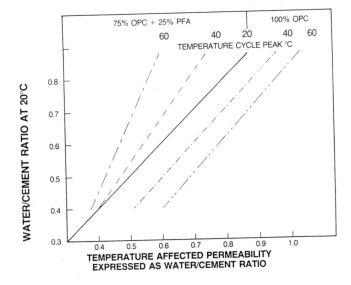

Fig 2 Effect of peak temperature on the permeability of concrete made with Portland cement, with and without pfa. (Owens 1985)

the water permeability of concrete subject to higher curing temperatures (Figure 2).

Whilst it has been usually accepted that higher temperatures degrade the performance of concretes made with all types of Portland cement, this degradation has usually been accepted as an inevitable consequence of their use.

Although composite cements, mainly based on blastfurnace slags, have been in use since about 1880, their wider use in hot climates has not until recently been accepted, mainly because the sources of granulated blastfurnace slag have not been readily available in hot climates, although this is not now the case. Similarly, the availability of the type of fly ash known as pulverised-fuel ash (pfa), has been confined mainly to the industrialised nations, but with the construction of power stations fuelled with hard coals, Cripwell (1992) identified that where these are pulverised and burnt in furnaces operating at about 1400°C provided a suitable secondary material with a glassy morphoplogy.

It is hoped from this brief assessment of the requirements and performance of concrete in hot climates, that the necessity of composite cements designed to accommodate the full range of resource materials, will become accepted standard practice for concrete specifications.

5 Components of Composite cements

ENV197-1 (1992) lists an acceptable "menu" of components for a range of common cements current in European use. In essence, the reactive materials listed as suitable for composite cements are:
 Portland cement clinker
 Granulated blastfurnace slag
 Pozzolanic materials
   Natural, or
   Industrial
 Fly ashes
   Siliceous, or
   Calcarious
 Burnt shale
 Silica fume
Amongst the more interesting materials are those listed as industrial pozzolanas, which are either thermally treated and activated clays and shales or air cooled slags from lead, copper, zinc and other products from the ferro alloys industries. While they should conform to the performance requirements of what is broadly described as "pozzolanicity", each material **ought to contain a high proportion of glass**, ie. say > 70%, in the form of reactive silica (SiO_2) and Alumina (Al_2O_3). However, whether all air cooled slags could ever have those oxides in the amorphous state, only their morphology can show, but granulation is the

most effective method of fixing the glass necessary for pozzolanicity or latent hydraulicity. It is this that is so critically important, as there have been so many failures, particularly with iron slags from open hearth processes which do not develop sufficient glass - **so beware.**

The testing techniques are also critically important as it is the detection, particularly of unstable materials such as **free lime**, which cause unsoundness or expansion in the concrete that have either to be elimimated or controlled. For instance, with calcareous fly ashes the "free lime" can vary from as low as 5% up to more than 20%, not only between sources, **but within sources**, so processing and homogenisation are essential aspects of that material's preparation and suitability,

Of the two most acceptable secondary materials available, are granulated blastfurnace slag and the siliceous types of fly ash arising from furnaces operated at above 1300°C. Of these two materials, when in sufficiently fine enough state, fly ash defined as pulverised-fuel ash, has a unique characteristic - a spherical particle shape which in itself helps to dewater concrete.

While it may not always be appropriate, pfa to BS3892 Pt 1 is at the high performance end of the spectrum of such materials, with the ability to minimise sulphate attack, alkali-silica reaction and delayed ettringite formation.

6 Assessment of secondary (hydraulic) materials

Standards are being developed for the application of secondary materials. The more recent revisions of standards such as BS3892 Pt 1 (1992) and BS6699 (1992), while defining the process control and requirements for the material, have also developed a unique procedure for the demonstration of equivalence of within-mixer combinations to appropriate composite cement standards containing those secondary materials. The assessment procedures have been designed for a combination of two materials which are manufactured, delivered and batched separately. There are, of course, necessary distinct differences from the procedures for combinations used for the sampling and testing of a factory blended composite cement.

However, the procedure for combinations determines the range of proportions over which their properties equate to the relevant composite cement standard. The versatility of the system is such that users may adopt any proportions within the declared range, and these need not be identical to the proportions used for testing. Troy (1992) describes the operation of such a procedure where authoritative standards exist and use both components for a composite cement. The challenge is the development of standards where there are available suitable materials such as those listed in ENV197-1 (1992), but where no standards exist for those materials in the processed form.

7 Test methods for assessing the suitability of available non-standard materials

Besides assessing the morphology of a material for its glass content, there is a number of test methods that give clear indications of any material's suitability, both chemically and physically.

Firstly, chemical influences, such as lime as CaO and magnesia as MgO, may be present in major quantities and it is whether they are present in the form of free lime or periclase that matters. The most appropriate and positive approach initially to determine their influence is by submitting the pulverised material to autoclave expansion tests to ASTM C-151 (1989). If satisfactory results are obtained to the limits of autoclave expansion and contraction to ASTM C-595 (1989), most of the other properties are controlled mainly by water requirement, heat of hydration and concrete soundness tests.

Water requirement is the next consideration and is very important, for if it increases proportionally with increasing percentage of secondary material, this is defeating the objective of lowering the water content of the concrete to below 150 litres/m^3. Sustaining, or even better, lowering the water requirement as the amount of secondary material increases, is a considerable attribute as it requires less admixture to effect the necessary reductions in water content of the concrete.

After water requirement, the assessment in concrete of the heat of hydration of a cement is necessary to measure the exothermic chemical reaction taking place. Here again the heat evolved depends upon the cement properties, the ambient temperature and the thermal characteristics of the test method. The temperature generated in concrete during hydration will depend upon the total heat evolved, the rate of heat evolution and the thermal efficiency of the conditions for casting the concrete. Livesey et al (1991) describe heat of hydration tests on a cement conducted by different methods and concluded that the Langavant semi-adiabatic calorimeter (1988) was a suitable method for the determination of the heat of hydration of a wide range of cements and proposes that for a "low heat" cement a maximum limit of 250 kJ/kg at 72 hours might be appropriate. The reactiveness of the secondary component can be clearly established by this technique and provided it also carries a minimum limit of 200 kJ/kg at 72 hours, this should be sufficient to control the temperature of concrete to below 8°C/100kg cement.

Finally, but due to its controversial nature, as it cannot identify whether the potential for deleterious expansion arises as a result of ASR or DEF, Duggan and Scott (1989) proposed a test that is nevertheless very useful for determining whether there is likely to be any delayed expansion problem in the concrete. If more than 0.05% expansion occurs within 20 days there is a risk of deleterious expansion from causes which have to be identified. The significance of this test is that it uses **all** the **actual materials** proposed. If expansion does occur, the causes are probably associated with DEF, and not ASR. The importance of this test is that it detects the effects of hot conditions, that is, as Heinz and Ludwig (1986) identified, is where the concrete temperature is likely to exceed 70°C as a result of temperature rise due to the combined effects of hydration or solar gain.

The greatest single influence, of course, is whether the hot conditions have influenced and benefitted the hydraulic reactivity of the secondary material, because temperature determines the relative proportions of the components and the later age strengths that are developed as a result of the hydraulic activity between appropriate components.

8 Curing Practice

Good curing practices for concrete made with composite cements, particularly in hot climates, are essential. Curing as a preference should be confined to the application of curing membranes in deference to other methods of curing. All free surfaces of freshly placed concrete should be treated with a pigmented solvent based curing membrane immediately any surface water has disappeared and just when the surface loses its moist sheen. For formed surfaces and immediately after removal of the formwork, a pigmented water based curing membrane should be applied.

The reason for requiring a pigmented curing compound is as a simple method of assessing compliance, and to see whether the specified curing practice has been undertaken! However, the performance of such curing compounds should at least have greater than 75% efficiency for 5 days. This should be sufficient under practical in situ conditions to obtain maximum hydraulicity of the composite cement and retain as much as possible of the lower mix water necessary to make the concrete as dense and non-permeable as possible.

9 Conclusions

The most suitable hydraulic cements for hot climates are composite cements made with Portland cement clinker and suitable (glassy) secondary components.

The cement should be capable of being dewatered with admixtures to give water contents lower than 150 litres/m³ in the concrete at temperatures up to 40°C, at the time of placing.

The assessment of suitable secondary materials relies on:

An amount greater than, say 70%, of glassy material present in that material.

The ability of the material to satisfy the autoclave expansion and contraction requirement of ASTM C-595.

Not to increase the standard water requirement of Portland cement with which it is used.

The heat of hydration, by semi-adiabatic calorimeter, to be not less than 200 kJ/kg nor greater than 250 kJ/kg at 72 hours.

The expansion of the concrete to be used and subjected to the Duggan Test should not be greater than 0.05%.

10 References

ASTM C151-86 (1991) **Standard Test Methods for Autoclave Expansion of Portland Cement.** Annual Book of ASTM Standards. Section 4 Construction. Volume 04.01. Cement; Lime; Gypsum. Philadelphia. pp. 121-124.

ASTM C595-89 (1991) **Standard Specification for Blended Hydraulic Cements.** Annual Book of ASTM Standards. Section 4. Construction. Volume 04.01. Cement; Lime; Gypsum. Philadelphia. pp. 290-294.

Association Française de Normalisation (NFP) 15-436 (1988) **Measurement of hydration heat of cements by means of semi-adiabatic calorimetry (Langavant Method).**

Bamforth, P.B. (1980) **In situ measurement of the effect of partial Portland cement replacement, using either fly ash (pfa) or a ground granulated blastfurnace slag, on the performance of mass concrete.** Proc Institute of Civil Engineers, London. Part 2 1980. 69 September. pp. 7-11.

Bamforth, P.B. (1986) **Alternative cements for hot climates.** Concrete. Journal of the Concrete Society. London. Volume 20 Nº 2. pp. 18-20.

British Standard BS3892 (1992) Pulverised-fuel ash, Part 1. **Specification for pulverised-fuel ash as a cementitious component for use with Portland cement.**

Cripwell, J.B. (1992) **What is pfa?** Concrete. The Concrete Society Journal. Wexham. May/June 1992. Volume 26. Nº 3. pp. 11-13.

Duggan, C.R. and Scott, J.F. (1989) **New test for deleterious expansion in concrete.** 8th International Conference on Alkali Aggregate Reaction. Kyoto. Japan. 17-20 July 1989

European PreStandard ENV197-1 (1992) **Cement - Composition, Specifications and Conformity Criteris - Part 1: Common Cements.** pp. 9-14.

Heinz, D. and Ludwig, U. (1986) **Mechanism of Secondary Ettringite Formation in Mortars and Concretes subjected to heat treatment.** ACI-SP 100-105. Concrete Durability. Proceedings of

the Katharine and Bryant Mather International Symposium.
Atlanta, USA. 1987. pp. 2059-2071.

Livesey, P., Donnelly, A. and Tomlinson, C. (1991) **Measurement of the heat of hydration of cement.** Paper presented to the International Conference on Blended Cements. University of Sheffield, England. 9-12 September 1991. pp. 28.

Owens, P.L. (1985) **Effect of temperature rise and fall on the strength and permeability of concrete made with and without fly ash in temperature effects in concrete.** ASTM STP 858 (ed. T.R. Nark). ASTM Philadelphia. 1985. pp. 134-149.

Richards, P.W. and Buttler, F.G. (1981) **The reaction of calcium hydroxide liberated on hydration of Portland cement with fly ash in mortars.** International Conference on Slags and Blended Cements. Mons. September 1981. pp. 7-11.

Troy, J.F. (1992) **Certification of combinations of Portland cement and pfa to BS6588.** Concrete. The Concrete Society Journal. Wexham. May/June 1992. Volume 26. N° 3. pp. 29-30.

18 PERFORMANCE EVALUATION OF CURING MEMBRANES

J. WANG
British Rail Research, Derby, UK
A. BLACK
Taylor Woodrow Management Ltd, Reading, UK

Abstract

Attempts were made in this study to evaluate the all-round performance of curing membranes. It was found that whilst the curing efficiency index, as specified in many standards and codes of practice, did correlate with the capability of the curing membranes in retaining moisture within concrete, such a single index is far from adequate and may be liable to give misleading information.

Keywords: Concrete, Curing Compound, Membrane Curing, Curing Efficiency, Moisture Loss

1 Introduction

Hot weather introduces problems in the manufacturing, placing, and curing of concrete. The rapid drying of fresh concrete associated with hot weather, for instance, may cause problems at both fresh and mature concrete stages. When concrete is hardening, rapid drying leads to excessive plastic shrinkage and, in severe cases, plastic shrinkage cracking. More importantly, its effect extends also into the hardened concrete stage, leading to reduced strength, reduced surface abrasion resistance, increased rate of carbonation and/or chloride ingress[1], all of which have serious implications with regard to the long-term durability of the concrete structures and their maintenance costs.

To prevent concrete, especially in hot climates, from an excessive rate of drying during the early period of its hydration, appropriate curing techniques have to be adopted, and one of the most frequently used is membrane curing. Upon application, the curing compound, often through the evaporation of its solvent, forms a membrane, which retards water evaporation from concrete.

There are two major families of curing compounds available, ie. solvent- and water-based types; between them some 40 different products are manufactured in the U.K. There is, however, no satisfactory method by which the likely performance of a curing membrane can be assessed. The existing British and American test methods are based on the concept of relative effectiveness of the membrane when

Concrete in Hot Climates. Edited by M. J. Walker. © RILEM
Published by E & F N Spon, 2 - 6 Boundary Row, London SE1 8HN. ISBN 0 419 18090 7.

applied on cement mortar specimens, and no correlation has yet been established between this efficiency index and its in-situ performance.

In this study, attempts were made to develop a method for comprehensive assessment of the performance of a curing membrane. Two curing membranes were selected, and they were tested initially for their curing efficiency index in accordance with a Draft British Standard. To simulate the real situation, concrete specimens were then made, onto which curing compounds were applied and the moisture loss from the concrete monitored. The effectiveness of the two curing compounds when applied on vertical surfaces, their degradability (under direct light), and their ability to provide protection against driving rain were also assessed.

2   Materials and Test Methods

2.1   Materials
For comparison purposes, one commonly used curing compound (F90) was used in addition to a specially formulated rain-erosion-resistant compound (H04). The general properties of these two curing compounds are given below in Table 1.

Table 1.   Details of curing compounds used

| Items | Product Code No | |
	H04	F90
Main component	1.1 trichloroethane resin /solvent	Resin/solvent
Colour	Reddish brown	Clear
Specific gravity (kg/m^3)	1.2	0.85
Curing efficiency* %	N/A	91
Manufacturer's recommended coverage rate, 1/m^2	0.07	0.20-0.25

* Determined in accordance with Department of Transport Specification for Road and Bridge Work Clause 2709, August 1977.

The cement mortar used for the curing efficiency test had a mix proportion of 1 : 3: 0.44 (cement:silica sand:water), as specified by British Standard Draft for Development 147[2]. For other tests, concrete was used and its mix proportion was 1 : 3.8 : 2.3 : 0.62 (cement :coarse aggregate:sand:water). Prior to their use, all the aggregates were dried at room temperature until they reached a stable moisture condition.

2.2 Test Methods

Six test methods were adopted and developed to measure the all-round performance of curing membranes, and they are briefly described below:

Touch dry: This is defined by the time for a curing compound, after its application, to reach non-stickiness upon touch. It varies from one product to another, and is usually within 2 to 3 hours.

Curing efficiency test: This is the test specified in BS DD147. Six mortar specimens, using the specified mix proportions, were prepared. Three of them were applied with a curing membrane at a rate of 0.20 ± 0.01 1/m², and the others were used as controls. The specimens were stored under an environment with a temperature of 38±1°C and a relative humidity of 35±5%. The curing efficiency of the curing membrane was then calculated from moisture losses at 3 days using the following formula:

$$\text{Curing efficiency (\%)} = \frac{(W_c - W_t)}{W_c} \times 100$$

where W_c - mean water loss from the control specimens(in %)
W_t - mean water loss from the test specimens (in %)

Moisture loss test: This was carried out using concrete specimens measuring 200 x 200 x 100 mm (depth). Curing compounds were applied 30 minutes after surface finishing to simulate typical site operations. Air-cured specimens were used as controls. All the specimens were stored in a controlled environment at 20 ± 1°C and 55 ± 5% R.H., and moisture loss from the specimens was monitored at 1, 3, 5 and 7 days respectively. The results were expressed as moisture loss per square metre of concrete surface, with allowance being made for the volatile components of the curing compounds.

Rain-erosion-resistance test: This test was designed to assess the rain-erosion resistance of curing membrane H04. The tests were carried out on site. Thirty minutes after a concrete slab (surface area approximately 1.5 m²) was cast, the curing compound was applied, at the manufacturer's recommended rate. Twenty minutes later, some three hundred millilitres of water were poured onto the slab from a height of 1.5 m in approximately 10 seconds. The resulting concrete surface was visually inspected and if no wash-out of the concrete occurred, the performance of the curing membrane was then considered to be satisfactory.

Applicability on vertical surface: This test was carried out by applying a curing compound, at a rate of some 0.15 1/m², onto newly demoulded vertical concrete surfaces, and then determining its retention. If more than 60% was retained, the curing compound was considered to be suitable for application on vertical surfaces.

Degradability: This was visually determined by the

colour change in the curing membrane when applied on a concrete surface and exposed outside over a period of 2 months. Many curing membranes are designed to breakdown and degrade in 1 month after application.

3 Results and Discussions

3.1 Curing efficiency test

The results of curing efficiency tests are shown below in Table 2. Each result represents the average of three individual tests. Both curing membranes are shown to have exhibited very high curing efficiencies, which all exceeded the requirement of 75% by BS8110 and the Department of Transport. The variation within each group of results was relatively small.

It needs to be remembered, however, that the tests were conducted at the same time using the same equipment. Between different laboratories, different results may well be expected, and such variations could sometimes be considerable[3,4].

Table 2. Results of curing efficiency tests to BS DD147

| Test | Curing Membrane | |
	H04	F90
1	92	89
2	96	93
3	94	90
Average	94	91

3.2 Moisture loss test

Moisture loss from concrete specimens was monitored up to 7 days, which was considered to be the most crucial period affecting the hydration of cement and hence the properties of concrete. The results are plotted in Figure 1 with each point representing the average of two specimens.

It is shown in Figure 1 that membrane curing significantly reduced moisture losses from concrete. Whilst a moisture loss of 3.41 $1/m^2$ was reached in air-cured specimens, specimens cured by H04 and F90 (both applied at the manufacturers' recommended rates) only reached 1.92 and 1.09 $1/m^2$, a reduction of 44 and 68% respectively. At an increased rate of application, the performance of membrane H04 was further enhanced, with a moisture loss of 0.78 $1/m^2$ at 7 days, a reduction of 77%. Since moisture loss occurs mostly in the surface region of the concrete at early ages, and it is the surface region of the concrete which has the greatest effect on the durability of a concrete element, the utilisation of membrane curing can therefore be expected to

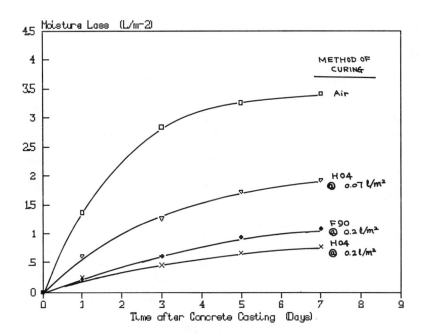

Figure 1. Effect of curing on moisture loss from concrete

enhance the performance of concrete significantly, through maximizing the degree of cement hydration.

The results given in Figure 1 show that both at manufacturers' recommended rates, curing membrane F90 is more efficient than curing membrane H04. This was thought to be due to the local discontinuities in the form of pin holes as observed in membrane H04 when applied at a low coverage rate. At an equal rate of application (ie. 0.2 l/m^2 as in the case of curing efficiency test), the performance of membrane H04 was better than that of membrane F90, and much better than itself at the recommended lower coverage rate.

This prompted the speculation that between the application rates of 0.2 and 0.07 l/m^2 (manufacturer's recommended rate), there must be an optimum figure when the maximum moisture retention can be achieved most economically. It has been suggested[5] that such an optimum figure relates to the water vapour permeability of an individual curing membrane, and is such that it produces a membrane having a vapour resistance of around 10×10^9 mm^{2}s mm Hg/g (vapour resistance = membrane thickness/its vapour permeability).

3.3 Touch Dry

For the curing membranes used, F90 reached touch-dry around 2 hours after application, but H04 did so in just 10-

15 minutes. The short touch-drying time of H04 is understood to be its unique feature, and is very important if it is to offer any early hour protection to wind-rippling and driving rain as claimed by the manufacturer.

3.4 Rain-Erosion Resistance
The rain-erosion-test was only carried out on curing membrane H04. It was observed that, where the water was poured, the concrete surface was dented slightly. However, the curing membrane still remained continuous and intact, and certainly no wash-out of the concrete occurred. This confirmed that the curing membrane was capable of protecting a fresh concrete surface against rain erosion, and was thought to be a very unique and beneficial feature of this product.

3.5 Applicability on vertical surfaces
It was observed that, after being applied onto vertical concrete surfaces, curing compound F90 almost immediately ran off. The retention ratio was estimated to be less than 20 to 30% of the applied liquid compound. However, most of H04 compound was seen to have been retained on the concrete surfaces after its application, and the retention ratio was well in excess of 80%. It was thought such different performances in these two curing compounds derived from their different viscosities, and clearly H04 was much more viscous than F90. The high rate of drying of H04 perhaps also contributed positively towards the its retention ratio.

The important point raised in this test is that, to achieve the desired effect of membrane curing on vertical surfaces, not only should the curing efficiency of the membrane be considered, but also the coverage which can be obtained. Without adequate coverage, the benefit of membrane curing cannot realised.

3.6 Degradability
As visual appearance of concrete surfaces is important in some cases, it is usually desirable that a curing membrane degrades after its designated function is accomplished, ie. as a moisture barrier.

Curing membrane H04 was non-degradable as stated by the manufacturer and this was confirmed from the site tests. Curing membrane F90 was described as degradable but, up to two months after its application, it was found to have degraded only slightly. Perhaps prolonged exposure would help to break down its composition. Both curing membranes can be expected to be worn off by traffic.

4 Conclusions

1) Both curing membranes exhibited similarly high curing efficiencies when tested in accordance with BS DD147, and

the variations within each group of results were small.

2) Membrane curing significantly reduced the moisture loss from concrete. When applied at the same rate, both membranes exhibited very high rates of moisture retention, with H04 being marginally more effective. However, when the coverage rate was reduced to the manufacturer's recommended level, the performance of H04 markedly deteriorated.

3) Membrane H04 reached touch-dry much more quickly than F90, and was found to be capable of protecting fresh concrete surface from "rain erosion".

4) It was found that curing compound H04 was suitable for application on vertical concrete surfaces but F90 was not. Both curing membranes were found to be non-degradable within the time span of the test (2 months).

5 Recommendations

1) A comprehensive evaluation of a curing compound should include, in addition to the curing efficiency test, the assessment of its application properties, touch-dry time, degradability and even perhaps its rain-erosion resistance.

2) Applicability of a curing compound on vertical surface should be positively stated in product specification to avoid misuse.

6 References

1) Wang, J. (1989) Membrane curing and performance of concrete, Ph.D Thesis, Dundee, U.K.
2) BS DD147 (1987) Methods of test for curing compounds for concrete, British Standards Institute.
3) Sparksman, W.G. (1978) An assessment of the standard tests for concrete curing membranes and consideration of a new test method. Project report for A.C.T Diploma Course, Institute of Concrete Technology, U.K.
4) Leitch, H.G. and Laycraf, N.E. (1971) Water retention efficiency of membrane curing compounds. Journal of Materials, JMSLA. Vol 6 pp 606-616.
5) Dhir, R.K., Levitt, M and Wang, J. (1989) Membrane curing of concrete: water vapour permeability of curing membranes, Magazine of Concrete Research, Vol 41, No 149, pp221-228.

7 Acknowledgement

The authors would like to thank Taywood Engineering Ltd. for its financial and technical assistance in carrying out this research work.

19 DURABILITY OF CONCRETE IN HOT CLIMATES: BENEFITS FROM PERMEABLE FORMWORK

W. F. PRICE and S. J. WIDDOWS
Taywood Engineering Ltd, London, UK

Abstract
The use of controlled permeability formwork (or CPF) is an effective way of improving the surface properties of concrete. Reductions in water absorption, carbonation, and chloride penetration can be demonstrated on concrete exposed to hot climatic conditions.

The improvements in surface quality resulting from using CPF are significantly greater than those produced by wet curing concrete cast against more conventional formwork. Indeed, using CPF may eliminate the need to apply any form of curing after removal of formwork.

The improved surface properties of concrete cast against CPF will reduce the penetration of the environment and consequently increase the durability of reinforced concrete structures.

The elimination of the need for curing is of practical significance in regions where climatic conditions make effective conventional curing difficult or impossible.

INTRODUCTION

Reinforced concrete structures in hot climates often deteriorate much more rapidly than structures in more temperate regions. In some cases, this accelerated deterioration is a result of the unavoidable use of poor quality or contaminated materials and the difficulties of achieving effective site curing of concrete structures exposed to hot environments are also well known. High initial concrete temperatures lead to rapid cement hydration, producing a weak porous microstructure of reduced strength and increased permeability (2) and loss of water from the concrete surfaces soon after removal of formwork is also detrimental to the long term durability of reinforced concrete structures.

Careful specification of the properties of permitted constituent materials, and limits on the overall mix proportions can be used to enhance the bulk properties of the concrete. However, the performance of the surface skin (the first line of defence against the penetration of deleterious agents such as water, sulphates, chlorides or carbon dioxide), can be significantly affected by the type of formwork used during construction and the type and duration (if any) of curing applied to the concrete after removing the formwork.

Considering the influence of formwork, Controlled Permeability Formwork (or CPF) allows water and air to pass through the fabric formwork liner while retaining cement. This leads to a reduction in near surface water/cement ratio with the associated enhancement of surface properties and an improvement in surface appearance as a result of the elimination of blowholes (3).

This paper examines the application of this technology specifically to concretes in hot climates, as a means of improving the durability of concrete structures. Two distinct categories of environment have been considered.

Concrete in Hot Climates. Edited by M. J. Walker. © RILEM
Published by E & F N Spon, 2 - 6 Boundary Row, London SE1 8HN. ISBN 0 419 18090 7.

a) Hot/Dry conditions, where high ambient temperatures are combined with low humidity. These conditions are typically found over much of the Middle East and in other arid regions throughout the world.

b) Hot/Wet conditions, where high humidity is combined with high ambient temperatures. This type of climate is often found in sub tropical regions (ie South East Asia).

A laboratory test programme has been undertaken to assess the effectiveness of the "Zemdrain" CPF system on durability related surface properties of concrete, specifically when used in hot climates. The formwork liner is based on the spunbonded polypropylene fabric, manufactured by Du Pont.

EXPERIMENTAL DETAILS

Concrete Mix Designs
The concrete mix used in this study was based on the use of OPC and natural gravel aggregates and sand. The mix proportions and compressive strength results are presented in Table 1. A mean (cube) strength of around 42 MPa was obtained (ie. typical of a grade C30 to C35 concrete).

Table 1. Details of concrete mixes (kg/m³)

	HOT/DRY CLIMATE	HOT/WET CLIMATE
OPC	325	325
20mm Gravel	805	805
10mm Gravel	405	405
Natural Sand	655	655
Water	185	185
W/C Ratio	0.57	0.57
Slump (mm)	120	120
Mix Temp (°C)	30	30
Mean 7 day compressive Strength (MPa) *	30.0	30.0
Mean 28 day compressive strength (MPa)	42.0	42.0

* mean of 3 standard cured 100mm cubes

Test Specimens
Concrete was batched by weight and mixed in $0.1m^3$ batches using a laboratory pan mixer. Test panels in the form of unreinforced concrete walls (750mm x 750mm x 150mm) were constructed, with one vertical face cast against an impermeable formwork (plywood coated with polyurethane varnish) and the opposite face cast against CPF.

The "Zemdrain" fabric was fixed to a plywood backing using adhesive tape around the perimeter. No drain holes were drilled in the plywood backing. However, the fabric extended below the base of the panel to allow for drainage. No mould release agent was required. In order to simulate concrete produced in a hot climate the sand and aggregate were stored in a warm cabinet overnight prior to mixing the concrete, and the mixing water was heated. An initial concrete mix temperature of 29-30°C was obtained.

The concrete was compacted into the moulds in three equal layers using an electronically powered 25mm diameter internal (poker) vibrator. During the placing and compacting process, and subsequently, clear bleed water was observed draining from the base of the CPF face of the mould. This demonstrated that the polypropylene fabric itself was essentially self-draining and confirmed that no drainage holes in the backing were necessary.

Immediately after compaction, the upper surface of the panel was covered in polythene sheeting, and the filled moulds were transferred to a preheated environmental chamber set at a temperature of 38 +/- 2°C.

The formwork was removed 24 hours after casting, and one vertical half of each panel (including both cast faces) was wrapped in wet hessian and polythene sheeting. The remaining half of the panel received no further treatment (no cure). One panel was then transferred to a second chamber under conditions of 38 +/- 2°C/90%min RH (water was prevented from coming into direct contact with the panels). The other panel remained at 38 +/- 2°C/50% RH.

The hessian and polythene (designated 'wet cure') were removed 3 days after casting the panels, and the panels then continued to be stored in their respective environmental chambers until they were 28 days old.

Thus for each set of climatic conditions, four curing regimes were obtained, ie.:

Impermeable formwork 'wet cure'
Impermeable formwork 'no cure'
CPF 'wet cure'
CPF 'no cure'

TEST METHODS

To avoid variability within the cast walls, resulting from the influence of settlement and water gain with height, all comparative sets of tests were undertaken at the same vertical location on each of the panels.

Following completion of the curing procedures described above, the following programme of tests was carried out.

Surface Hardness
Measurements of surface hardness (or rebound number) were made on the concrete using a Schmidt Hammer in accordance with BS 1881 (4). A minimum of twelve individual readings were made for each curing condition. No attempt was made to correlate the rebound number measurements with the compressive strength of the concrete.

Initial Surface Absorption
The initial surface absorption rate was measured using the ISAT test after conditioning the panles at 20°C/50% RH for a minimum of 48 hours (5).

The absorption rate was measured at 10 min, 30 min, 60 min, and 120 min, and a cumulative 2 hour absorption was also calculated (3).

Chloride Ingress
A 100mm diameter core was cut from the panels for each combination of climatic condition and curing regime.

The cores were then vacuum saturated in distilled water until the increase in weight was less than 0.5g per kg of sample, per day. Following the completion of vacuum saturation, the cores were immersed in 5M NaCl solution saturated with $Ca(OH)_2$ and maintained at 40°C. After 28 days the cores were removed from the solution and rinsed with water to remove excess saline solution from the surface of the sample.

The sample was then sliced vertically using a minimal amount of water lubricant. One half of the sample was dried at 105°C and then cooled to room temperature in a dessicator over silica gel.

The dried sample was mounted in a precision vertical milling machine and the test face ground using a diamond impregnated flat steel tool rotating in a plane parallel to the test face.

The sample was ground in increments of 1 or 2mm depth from the cast face (checked using the integral micrometer on the milling machine). Only the central portion of the face was ground, thus avoiding contamination from lateral chloride penetration.

Powder from each of the depth increments was collected and passed through a 150 micron sieve prior to analysis. The chloride content of each increment was then determined by potentiometric titration.

Accelerated Carbonation

Cores were removed from the test panels, and all surfaces of the cores apart from the cast faces were sealed in polythene sheet and heavy duty adhesive tape. The cores were then stored in an environment consisting of air with an enriched carbon dioxide content of 4% to accelerate carbonation, a relative humidity of 50% and a temperature of 23 +/- 2°C. Dhir et al (6, 7) have stressed the importance of maintaining an appropriate relative humidity in the promotion of carbonation.

After continuous storage in the carbonation chamber for a period of 12 weeks, the cores were split open and sprayed with a phenolpthalein indicator solution. The depth of carbonation from the cast face was then measured 24 hours later.

DISCUSSION OF RESULTS

The results of the tests on the surface properties of concrete exposed to simulated hot climates, and cast against different formwork types are presented in Table 2.

Surface Hardness
The 28 day surface hardness measurements are shown in Figure 1.

The test panels show certain interesting features.

1) All surfaces cast against CPF were markedly improved.

2) The improvement was most apparent for the concrete stored in Hot/Wet conditions, with results being consistent with those obtained in earlier studies (3).

3) The enhancement produced in Hot/Dry storage conditions was much less significant and less consistent with results from earlier studies.

4) Very little difference was observed between any of the surfaces cast against impermeable formwork, regardless of curing or storage conditions.

The final observation, 4) was somewhat surprising.

However, subsequent more detailed examination of these test panels, indicated that there had been a rapid carbonation of the surfaces cast against conventional impermeable formwork (measured on cores sprayed with phenolpthalein) when compared with the CPF faces (Table 3). Similar early carbonation has been observed in exposure trials in the Arabian Gulf (8).

It is possible therefore, that it is the early precipitation of calcium carbonate in the near surface pore system of concrete cast against conventional impermeable formwork, which had produced a denser (harder) surface, and that this has offset the detrimental effect of storage in a dry environment.

This also explains the apparent reduction in the benefit if CPF for the Hot/Dry conditions, having been compared with a surface with a hardness enhanced by carbonation.

210

Table 2. Surface properties of concretes exposed to simulated hot climates
(28 day results).

PROPERTY	UNITS	CLIMATE	FORMWORK TYPE	CURE	RESULT
28 DAY SURFACE HARDNESS	-	Hot/Dry	Imp	Wet	34.3
			Imp	None	32.4
			CPF	Wet	38.6
			CPF	None	37.2
		Hot/Wet	Imp	Wet	33.9
			Imp	None	33.0
			CPF	Wet	45.6
			CPF	None	44.6
10 MINUTE ISAT	ml/m²/s	Hot/Dry	Imp	Wet	0.46
			Imp	None	0.55
			CPF	Wet	0.27
			CPF	None	0.26
		Hot/Wet	Imp	Wet	0.03
			Imp	None	0.07
			CPF	Wet	0.002
			CPF	None	0.002
2 HOUR CUMULATIVE ISAT	m/m²	Hot/Dry	Imp	Wet	1490
			Imp	None	1643
			CPF	Wet	788
			CPF	None	779
		Hot/Wet	Imp	Wet	99
			Imp	None	266
			CPF	Wet	3
			CPF	None	11
ACCELERATED CARBONATION DEPTH	mm	Hot/Dry	Imp	Wet	13.5
			Imp	None	14.5
			CPF	Wet	4.5
			CPF	None	6.0
		Hot/Wet	Imp	Wet	9.5
			Imp	None	10.0
			CPF	Wet	<1
			CPF	None	<1

Fig. 1. Effects of CPF on the surface hardness of
concrete measured at 28 days.

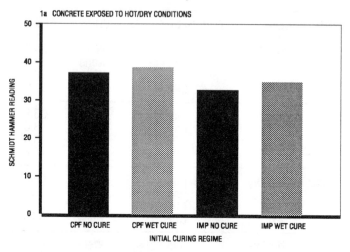

1a CONCRETE EXPOSED TO HOT/DRY CONDITIONS

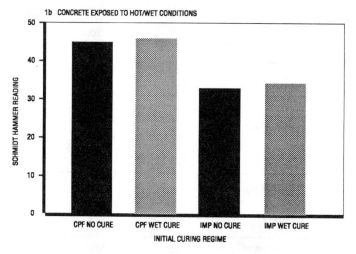

1b CONCRETE EXPOSED TO HOT/WET CONDITIONS

Nevertheless, for all conditions of curing and storage, the enhancement of surface hardness achieved by casting concrete against CPF was substantial and much more significant than the improvements produced by curing alone.

Although surface hardness per se, is not a property which normally merits much consideration with regard the performance of concrete structures, its relationship to compressive strength and hence water/cement ratio (albeit qualitative), makes it a useful indicator of general surface quality and warrants its inclusion in this discussion.

Table 3. Early age natural carbonation of concrete surfaces exposed to simulated hot climates (carbonation depths measured 17 weeks after casting).

CLIMATE	FORMWORK TYPE	CURE	CARBONATION DEPTH (mm)
Hot/Dry	Imp	Wet	3 - 4
	Imp	None	5 - 6
	CPF	Wet	< 1
	CPF	None	< 1
Hot/Wet	Imp	Wet	< 1
	Imp	None	< 1
	CPF	Wet	0
	CPF	None	0

Surface Absorption

The absorption of water, and substances dissolved within it, through the surface of a concrete structure, plays a major role in all of the principal deterioration mechanisms. The ISAT test has been shown to be a sensitive indicator of the influence of concrete materials, mix proportions, and curing, on the surface properties of concrete (3, 9, 10).

Examining the data obtained for concrete exposed to Hot/Dry conditions, it is apparent that the levels of 10 minute absorption rate were all relatively high compared with values for surfaces cast against CPF. When assessed in relation to proposed Concrete Society (11) performance levels of 0.25 to 0.50ml/m^2/sec for typical structural concrete, the surface cast against impermeable formwork with no curing would be considered to exhibit 'high absorption', while the same surface subjected to a 3 day wet cure exhibits absorption in the 'average' category. The rapid surface carbonation observed in concrete cast against impermeable concrete does not appear to reduce surface absorption, probably as a result of the shrinkage associated with carbonation (12) leading to increased surface porosity. Using CPF further reduced the surface absorption to the low end of the 'average' range.

The influence of curing following the removal of the impermeable formwork is clearly apparent, demonstrating the need for effective curing in dry conditions. However, when concrete was cast against CPF, not only was the influence of curing virtually eliminated, but also the reductions in surface absorption were much greater than could be achieved by wet curing alone.

In the Hot/Wet exposure conditions, the overall level of ISAT was considerably reduced for all combinations of formwork and curing. This is believed to be a consequence of the higher humidity which not only aids continued hydration, but also maintains the concrete at a higher level of saturation. Even here, however, the influence of CPF was apparent, producing a surface with almost no initial surface absorption, independent of cure.

It is clear then, that casting concrete against CPF formwork provides a means of reducing the water absorption through the concrete surface in both Hot/Wet and Hot/Dry conditions. This would be expected to improve the durability of concrete structures, particularly in arid conditions, where overall levels of absorption are often very high. Using CPF also considerably reduces the need for curing (often very difficult to achieve effectively in dry climates), offering a practical possibility to achieve high quality durable concrete surfaces under difficult conditions.

213

Fig. 2. Effects of CPF on the initial surface
absorption of concrete at 28 days.

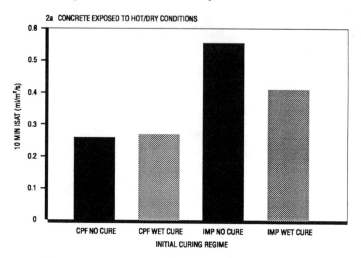

2a CONCRETE EXPOSED TO HOT/DRY CONDITIONS

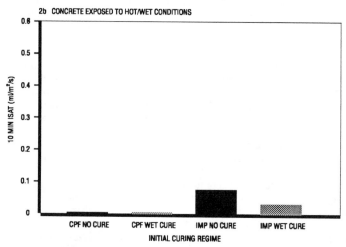

2b CONCRETE EXPOSED TO HOT/WET CONDITIONS

Chloride Ingress

Penetration of chlorides through the cover zone of concrete structures can lead to the corrosion of embedded reinforcement (1) which is probably the major cause of deterioration of reinforced concrete structures worldwide.

Chloride penetration is though to be influenced by both the initial absoption (through its effect of building up high surface chloride concentrations) and a longer term diffusion process. Chloride levels at the depth of the reinforcement will increase until steel corrosion is activated at a certain threshold level of chloride (often taken as 0.4% Cl by weight of cement).

The penetration of chlorides into concrete is often considered to be represented by Ficks second law of diffusion (13, 14), a solution to which is given below:-

Fig. 3. Chloride ingress profiles in concrete
(concrete 28 days old at exposure to chlorides).

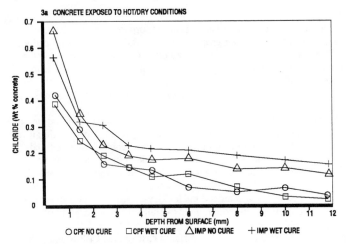

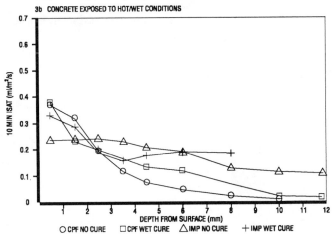

$$Cx = Cs \left(1 - erf \frac{x}{2\sqrt{Dt}}\right)$$

where

Cx	=	Chloride concentration at depth x
Cs	=	Surface chloride concentration
D	=	Diffusion coefficient
t	=	Exposure tme
x	=	Depth
erf	=	Error function

The build up of chloride at depth is thus controlled by two major factors, namely the diffusion coefficient of the concrete and the surface chloride concentration.

Surface Chloride Levels

The surface concentration is influenced by the initial adsorption/absorption characteristics of the concrete (15, 16) which are in turn affected by mix proportions, materials and curing procedures. Casting concrete against CPF has already been shown to be an effective means of reducing the surface absorption of concrete, and in an earlier study (3) a reduction of the chloride diffusion coefficient in the near surface zone of concrete cast against CPF, was measured.

Examination of the chloride ingress profiles for concrete exposed to Hot/Dry conditions, shows a number of interesting features (Fig 3a).

The influence of both formwork type and initial curing on the surface chloride concentration is evident. Concretes cast against impermeable formwork exhibited a higher surface chloride level which was reduced by initial wet curing. In contrast, the concrete cast against CPF exhibited lower surface chloride levels and appeared to be less influenced by the initial curing regime.

When concrete is exposed to Hot/Wet conditions, certain differences are observed (Fig 3b). The overall level of surface chlorides for all combinations of formwork type and curing regime were much closer together and generally lower than achieved in Hot/Dry conditions. This to some extent reflects the influence of higher humidity on reducing the overall surface absorption of all the concretes as demonstrated by the ISAT results.

Under the wet storage condition the levels of chloride in surfaces cast against CPF were slightly higher than those cast against conventional formwork. This is thought to result from the higher surface cement content of the concrete cast against CPF, which has an associated ability to hold

Fig. 4. Accelerated carbonation penetration of concrete.

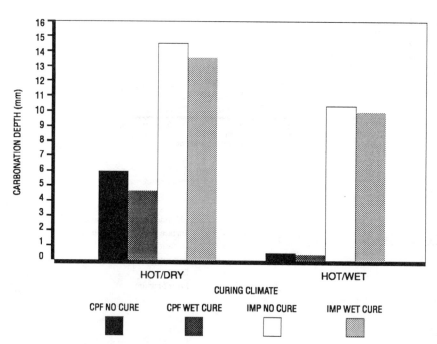

(absorb) high levels of Chlorides (17). This acts in opposition to the effects of CPF in reducing surface absorption.

Chloride Diffusion Coefficient

The level of chloride at depth (up to 12mm) is clearly influenced by the type of formwork used, with CPF producing a lower level, independent of climate and initial curing.

To compare results it is often convenient to make an estimate of chloride diffusion coefficient (D) by fitting a curve to this data and applying equation 1. A number of authors (14, 15) have argued that, because of the difficulty in ensuring steady state conditions in chloride ingress tests, caution should be used in estimating true diffusion coefficients, the term 'effective diffusion coefficient' (D_e) being more appropriate. Values for D_e are presented in Table 4 calculated on the assumption that the level of chloride measured in the 0-1mm increment represents C_s and that the curve passes through the measured chloride content at 8mm depth.

These results demonstrate that the effective chloride diffusion coefficient can be reduced by casting concrete against CPF. The magnitude of this reduction varies from around 50% in Hot/Dry conditions to an order of magnitude in Hot/Wet conditions.

The reduction in the rate of build up of chlorides at reinforcement depth resulting from the use of CPF significantly reduces the risk of chloride induced reinforcement corrosion in hot climates.

Table 4. Chloride ingress parameters for concretes in hot climates (calculated from measured chloride ingress profiles).

CLIMATE	FORMWORK TYPE	CURE	SURFACE CHLORIDE LEVEL (% concrete) C_s	EFFECTIVE DIFFUSION COEFFICIENT ($m^2/s \times 10^{-12}$)D_e
Hot/Dry	Imp	Wet	0.56	12.7
	Imp	None	0.67	7.8
	CPF	Wet	0.39	6.5
	CPF	None	0.42	4.7
Hot/Wet	Imp	Wet	0.32	34.2
	Imp	None	0.24	35.8
	CPF	Wet	0.35	8.2
	CPF	None	0.35	3.6

CARBONATION

In certain areas of the world (eg Hong Kong and Bahrein), carbonation is a major cause of reinforcement corrosion (9, 18). The reduction in pH, associated with the reaction between atmospheric carbon dioxide and calcium hydroxide in the cement paste, causes a depassivation of the surface of the embedded reinforcing steel. Subsequent ingress of moisture (and/or chlorides) will cause rapid corrosion.

The results of the accelerated carbonation tests are shown in Table 2 and Figure 4. It is clear that carbonation occurred most rapidly in the Hot/Dry climate, but even under these adverse conditions, the use of CPF more than halved the depth of carbonation, even when no curing was subsequently applied. This effect is more significant than the improvement made by wet curing.

In the Hot/Wet climate, the carbonation depth in concrete cast against impermeable formwork was around 30% lower than in the dry climate. The effect of CPF was almost to eliminate carbonation.

In both climatic conditions, the effect of CPF alone on the reduction of carbonation depth (independent of cure) was more significant than the application of wet curing.

CONCLUSIONS

When compared with the use of conventional impermeable formwork the use of controlled permeability formwork (CPF) produced improvements in the surface properties of concrete cast and stored in simulated hot climates. In particular, the following conclusions were reached.

a) The surface hardness of concrete cast against CPF (itself a qualitative measure of 'surface quality') was increased. The improvement was most marked for concretes exposed to a Hot/Wet environment. It is believed that rapid surface carbonation tended to enhance the otherwise poor surface hardness of concrete cast against impermeable formwork when exposed to Hot/Dry conditions, thus diminishing the apparent benefits of CPF under these conditions.

b) The surface absorption of concrete was reduced by CPF in both Hot/Wet and Hot/Dry conditions, and concrete cast against CPF was less sensitive to curing.

c) The effective chloride diffusion coefficient of concrete cast against CPF was reduced by about 50% in Hot/Dry conditions and by as much as an order of magnitude in Hot/Wet conditions. The effect of CPF on chloride build up at depth (independent of cure or exposure conditions), was always more significant than applying a 3 day wet cure to concrete cast against impermeable formwork.

d) CPF caused a reduction in the carbonation of concrete (measured using an accelerated test method), of about 9mm in both Hot/Wet conditions (5mm vs 14mm) and Hot/Dry conditions (10mm vs 1mm). In the latter case, carbonation was virtually eliminated. The effect of curing on reduction of carbonation was insignificant compared with the effect of CPF.

e) The overall effect of using CPF, was to enhance consistently the surface properties of concrete in both Hot/Wet and Hot/Dry conditions. The effects of curing were less significant than the effects of CPF alone, with the consequence that using CPF desensitizes the concretes against the effects of inadequate curing.

This has significant practical advantages for producing durable reinforced concrete structures in hot regions where climatic conditions are such that effective curing is often extremely difficult. This is probably the major practical benefit for the use of CPF in hot climates.

ACKNOWLEDGEMENTS

The authors wish to thank the directors of Taywood Engineering Limited for permission to publish this paper. The financial contribution of Du Pont de Nemours (Luxembourg) SA in support of this work is acknowledged.

REFERENCES

1. Treadaway K.W.J., **A review of comparative performance of reinforced concrete in UK and Arabian Gulf Conditions.** Proc. Ist Int. Conf. Deterioration and repair of reinforced concrete in the Arabian Gulf, Bahrein Society of Engineers, October 1985, vol 1 301 - 308

2. Verbeck G .J. & Helmuth R.H., **Structure and physical properties of cement paste.** Proc. 5th Int. Symp. Chemistry of Cement, Tokyo 1968, vol. 3, 1 -32.

3. Price W.F. & Widdows S.J., **The effect of permeable formwork on the surface properties of concrete,** Magazine of Concrete Research, vol 43 no. 155, 1991, 93 - 104.

4. BRITISH STANDARDS INSTITUTION, **BS 1881: Methods of Testing Concrete, Part 202.** Recommendations for surface hardness testing by rebound hammer, London 1986.

5. BRITISH STANDARDS INSTITUTION, **BS 1881: Methods of testing Concrete, Part 5, Method 6.** Test for determining the initial surface absorption of concrete, London 1970.

6. Dhir R.K., Hewlett P.C. & Chan Y.N., **Near surface characteristics of concrete: Prediction of carbonation resistance.** Magazine Concrete Research vol. 41 no. 148, 1989, 137 - 143.

7. Dhir R.K., Jones M.R. & Munday K.W.J., **A practical approach to studying carbonation of concrete.** Concrete vol. 19 no. 10 1985, 32 - 34.

8. Macmillan G.L. & Treadaway K.W.J., **An exposure trial of concrete durability in Arabian Gulf conditions.** Proc. 3rd Symposium, Corrosion of reinforcement in concrete construction, SCI 21 - 24 May 1990, 109 - 118.

9. Dhir R.K., Hewlett P.C. & Chan Y.N., **Near surface characteristics of concrete: Assessment and development of in-situ test methods.** Magazine Concrete Research vol. 39 no. 141 1987, 183 - 195.

10. Dhir R.K. & Byers E.A., **PFA concrete near surface absorption properties.** Magazine Concrete Research vol. 43 no. 157 1991, 219 - 232.

11. THE CONCRETE SOCIETY, **Permeability testing of site concrete: A review of methods and experience.** Technical Report No. 31, London 1988.

12. F.M. Lea, **The chemistry of cement and concrete 3rd edition,** Edward Arnold, London 1970.

13. Miyagawa T., **Durability design and repair of concrete structures: Chloride corrosion of reinforcing steel and alkali-aggregate reaction.** Magazine Concrete Research vol. 43 no. 156, 1991, 155 - 170.

14. Dhir R.K., Jones M.R. & Ahmed H.E.H, **Concrete Durability: Estimation of chloride con centration during design life.** Magazine Concrete Research vol. 43 no. 154, 1991, 37 - 44.

15. Bamforth P.B. & Pocock D.C., **Minimising the risk of chloride induced corrosion by selection of concreting materials.** Proc. 3rd. Int. Symp. Corrosion of reinforcement in concrete construction, SCI, 21 - 24 May 1990, 119 - 131.

16. McCarter W.J., Ezirim H. & Emerson M. **Absorption of water and chloride into concrete.** Magazine Concrete Research vol. 44 no. 158, 1992, 31 - 37.

17. D.C. Pocock & P.B. Bamforth, **Cost effective cures for reinforcement corrosion - collaborative research brings new developments.** Construction Maintenance & Repair, Nov/Dec 1991, 7 - 10.

18. Poon C.S. & Baldwin G.R., **Corrosion of steel in concrete - A basic understanding and research needs in Hong Kong.** Hong Kong Engineer, July 1989, 22 - 28.

20 THE MANUFACTURE OF PRECAST CONCRETE COUNTERWEIGHT UNITS FOR THE SUKKUR BARRAGE, PAKISTAN

C. R. ECOB and D. S. LEEK
Special Services Division, Mott MacDonald Ltd, Croydon, UK
B. VOAK
Whessoe Projects, Darlington. Co. Durham, UK

Abstract
This paper describes the specification, mix design and production trials and the manufacture of reinforced concrete, gate counterweight units for the Sukkur barrage on the River Indus in Pakistan. It details the features of concrete specification, selection, storage and pre-conditioning of raw materials, the procedures used during mixing and casting and the curing regime for the concrete units implemented to minimise the effects of producing concrete in a hot arid climate. The production of the units showed that by taking adequate precautions during all stages of the production process, it was possible to produce concrete of high quality using only the intermediate technology available locally.

1 Introduction

The Sukkur barrage on the River Indus in Sindh Province, Pakistan, is an impressive structure consisting of masonry piers and reinforced concrete arches (see Figure 1), approximately 1.5km and having 66 arched gate bays. Shortly after the barrage's golden jubilee celebrations in 1982, a corroded steel gate near the middle of the river collapsed. Only swift action in opening all the other gates to lower the river level averted a major disaster. The implications of what could have happened lead to a major seven year rehabilitation programme due to end in 1993.

The barrage is vital to the agriculture of Sindh province, controlling water supply from the Indus to over 32 million hectares of farmland and forms a key component in the world's largest irrigation scheme. Fortunately the failure occurred during the winter months when flows were only about 1000 cumecs. During the summer, the flows reach approximately 34,000 cumecs. A failure of a gate during peak flow would induce vibration in the massive barrage structure and possible result in catastrophic failure, not just of the barrage, but for much of Sindh province. The Government of Sindh appointed international consulting engineers Mott MacDonald to recommend long term remedial measures.

2 Investigation and remedial options

A full inspection of the steel gates was carried out in 1984 during which it was found that virtually all gates were suffering from corrosion, with up to a 30% loss in steel cross section. Consequently it was recommended that all gates be replaced, using a technique permitting replacement during the low flow winter months, with minimum

Concrete in Hot Climates. Edited by M. J. Walker. © RILEM
Published by E & F N Spon, 2 - 6 Boundary Row, London SE1 8HN. ISBN 0 419 18090 7.

Fig. 1. Sukkur Barrage

disruption to the operation of the barrage. To achieve this a caisson gate, suspended from a pontoon, was floated into the gate bay and attached to the pier structures. In order to establish whether the piers had sufficient capacity to support the new (approximate) 200 tonne load, to each, a major inspection and pier strengthening exercise was undertaken.

The pier strengthening and original steel gate replacement formed the major components of the rehabilitation. Following the commencement of the Contract it was decided to investigate the condition of the gate counterbalance units, which were contemporary with the gates and consisted of riveted steel plate boxes containing rock ballast. Although appearing in good condition, they did exhibit small areas of rust staining. Several units were removed for detailed examination during the gate replacement programme and following removal of the ballast, significant internal corrosion was found. The condition in the units where chlorides were present, were conducive to slow rates of corrosion. It was decided that it would be cost effective to replace the counterweight units during the gate replacement contract, being undertaken by the UK contractor, Whessoe Projects Ltd.

There was considerable debate concerning the type of material from which the new units should be manufactured; The client preferring a similar ballasted steel box which had been seen to have performed adequately for 50 years, which with more sophisticated protective systems now available, could probably perform adequately for much longer. Mott MacDonald recommended a reinforced concrete alternative, that would both be durable and more economic. With the current advances in concrete technology, it should be possible to achieve an equivalent service life to the steel alternative, with a lower maintenance cost.

The Client, The Irrigation and Power Department of the Government of Sindh, accepted the proposed reinforced concrete alternative, but emphasised the importance of durability and good visual appearance at close proximity, owing to the close proximity of vehicular and pedestrian traffic.

3 Concrete performance requirements

An inspection of concrete structures, principally bridges, in the vicinity of Sukkur gave an indication of the quality of local concrete technology and that to achieve the required standard for the counterweight units would necessitate the use of overseas expertise. Although an earlier inspection of the *insitu* concrete components of the barrage itself, the piers and arches, showed it to be in fair condition, carbonation depths of up to 100mm and chloride levels (by weight of cement) of up to 3% were recorded. Corrosion of reinforcement was not extensive probably due to the very low relative humidity that existed for much of the year although the chloride concentration clearly indicated the presence of relatively high levels of air borne salt from the surrounding salt flats. It was therefore essential that a dense, low permeability concrete be produced. A requirement by the client that all plant and materials be sourced locally increased the already difficult challenge of producing such a high quality concrete.

The Contractor whose principal experience and expertise was in steel fabrication and erection, had the option of appointing a specialist precast concrete subcontractor, but elected to undertake production himself. This approach caused some concern to the Consultant before production started, although there was a requirement in the contract bill of quantities for an experienced expatriate concrete technologist and laboratory supervisors. Production also involved the recruitment and training of local labour with no knowledge of working conditions and with little or no education. However, by limiting individual groups of workers to fixed tasks, with close supervision it was possible to produce, in a repeatable manner, the required quality of concrete.

A major factor in the manufacture of the 64, 16.3m long units, comprising of 704 precast components, were the very high ambient temperatures found at Sukkur, one of the hottest regions of Pakistan. Temperatures above forty degrees centigrade can be experienced for 8 months of the year and as high as 50°C occasionally reached. The construction programme which essentially dictated a relatively slow rate of production, coupled with the scale of precasting operation, that would be forced on the Contractor, ensured that casting would be undertaken through all periods of the year.

There are two periods when problems occur when producing concrete in high ambient temperatures (above 30°C), during production i.e. mixing, placing and during the concrete hardening processes.

3.1 Concrete Production
High temperatures result in a decrease in workability during placing, with the reduction of slump being related to the ambient temperature at the time of mixing. This also increases the rate of loss of workability which necessitates minimising the interval between mixing and placing. There is also an increased risk of plastic shrinkage cracking due to rapid evaporative water loss from the surface of the concrete.

3.2 Concrete Hardening
Heat generated during cement hydration causes the concrete temperature to increase. This is followed by cooling and associated contraction, which, if unrestrained will not result in cracking. In practice the surfaces of the cast element cool faster than the core, producing a differential temperature and therefore stress gradient between the core and the surface layers which results in cracking.

3.3 Techniques to Lower the Initial Concrete Temperature

It is essential to keep the wet concrete temperature below 30°C in order to improve concrete workability and prevent the maximum temperature becoming too high. The principal methods of keeping fresh concrete cool available on site were:

Cement	- Air conditioned storage of bagged cement.
	- Hot cement (>60°C) not to be used.
	- New cement to be cooled before use
Mix water	- Addition of crushed ice.
	- Insulated storage containers.
Aggregates	- Protect from solar heating.
	- Cooling coarse aggregate with water.

4 Constituent Materials

The Client required that the raw materials for the production of the concrete units be, where possible sourced locally. This lead to the following materials being used.

4.1 Cement

OPC from the local Rohri factory was produced to BS12 : 1958. The chemistry showed a C_3A level of between 10.9% and 12.0%, C_3S level between 42% and 45% and C_2S level between 27% and 29%. Such cement would exhibit quite good early strength, low mid-age strength, but high long-term strength. There was initial concern that with the cement composition above, the time required to reach the lifting strength of the precast components ($25N/mm^2$) would be extremely long irrespective of the quantity of cement used.

4.2 Aggregate

The coarse aggregate was a bituminous dolomitic limestone from the river bed deposits at Machh Dhadar. It was a high density, coherent rock which complied with the requirements of BS 882. When tested for alkali reactivity it was found not to be reactive.

The fine aggregate (Bolhari sand) consisted of limestone and quartz with minor amounts of chert. It was also found not to be alkali reactive.

4.3 Water

The water used in the production of concrete can have a significant effect on the quality, depending on the impurities present. Potable water was not available, so water from tube wells sunk on the site, which was tested for compliance with BS 3148, was used for concrete production.

4.4 Admixtures

An imported retarder/plasticiser, which complied with the requirements of BS 5075, was used in the concrete to increase workability and placing time for the concrete.

4.5 Steel reinforcement

Reinforcing steel was also imported. This was shipped to Pakistan as deck cargo which on arrival at the site was found to have been contaminated with salt. Before use all the reinforcement was blast cleaned to remove the contamination.

5 Specification

The specification required the Contractor to maintain the ambient temperature of the mixed concrete to between 16° and 30°C. The methods to achieve this requirement were to be proposed by the Contractor in his working procedure, for approval by the Engineer. The specification also prohibited the placing of concrete at air temperatures in excess of 30°C, without the specific approval of the Engineer. All production was to take place in a specially design facility, shaded from direct sunlight.

A series of trial mixes was specified to develop both the concrete mix and the procedures to be used during casting.

6 Proposed working procedure

The problems associated with concreting at high temperatures were itemised in the specification. The practices for overcoming these problems were identified and suitable procedures adopted and established before and during production and by the site trials.

6.1 Cement
'Fresh', hot cement, direct from the works was not allowed to be used. Cooling to below 25°C was specified as was stock rotation.

6.2 Aggregate
It was not possible to make effective reductions in the temperature of mixed concrete by attempting to artificially cool the fine aggregate, only shading from direct sunlight was therefore specified. The coarse aggregate which supplied a significantly greater quantity of heat to the concrete was to be both shaded and cooled (by continuous spraying with water). The additional moisture content of the aggregate was taken into account when calculating the water/cement ratio of the concrete.

6.3 Water
The mixing water was to be cooled and stored in a shaded, white painted tank prior to use. Ice (produced from the borehole water) was to be added to the mixing water 5 to 6 hours prior to mixing to achieve a maximum temperature of 16°C during production.

6.4 Concrete
The temperature of the fresh concrete immediately after mixing was to be in the range 16° to 30°C, with the intention being to achieve the lower value. Production was to be timetabled to consider the ambient temperature conditions to achieve the temperature requirements of the concrete.

The concrete mix was designed to achieve the strength requirements, while minimising the quantity of cement, consequently minimising the temperature rise associated with the hydration reactions.

6.5 Curing
A rigorous 12 day curing period was specified. This was carried out in three stages, after casting.

1. With the shutters still in place, the exposed concrete surfaces were to be covered with polyethylene sheet until initial set occurred.
2. The exposed top surface of the unit was then to be covered with continuously wetted hessian (wetting was achieved by means of a perforated hose), beneath a

specially designed white, full unit, polyethylene cover to provide both shade and a high relative humidity environment.

3.　After removal of the shutters (after a minimum of 24 hours, but timetabled for the coolest part of the day), the complete surface of the concrete was to be covered with wetted hessian and the cover replaced. The units were to be left in this environment for a further 11 days.

6.6 Quality Assurance

The Contractors quality plan required that the temperatures of all the constituents of the concrete be measured and recorded immediately before mixing and also that the temperature (and slump) of the mixed concrete and the ambient temperature be taken and recorded.

7 Concrete mix trials

The concrete mix design was developed in three stages; preliminary mix development, to produce a base mix and to establish the storage requirements for the concrete constituents, final mixing trials, to refine the selected base mix into the production concrete and casting trial panels (with representative steel reinforcement), using the procedures expected to be used during production, to be used as standards for acceptance of the surface finish.

7.1 preliminary mix development

Four mix designs were initially trialed as shown in Table 1. The concrete was mixed in a 25kg mixer with all the constituents weigh batched except the admixture. The temperature of all the materials was measured and 150mm cubes were made for strength testing. The strength characteristics of the trial mixes are summarised in Table 2.

Table 1. Initial trial mix designs

Ref	Cement (kg)	Aggregate Coarse (kg)	Fine (kg)	Fine (%)	Free w/c	Admixture (l)
TM1	330	1055	895	46	0.39	2
TM2	350	1045	835	44	0.40	2
TM2A	350	1055	875	45	0.41	2
TM3	370	1055	855	45	0.37	2

Table 2. Compressive strength results

Ref	Slump (mm)	3 day (N/mm^2)	7 day (N/mm^2)	14 day (N/mm^2)	28 day (N/mm^2)
TM1	85	30.0	34.5	36.5	39.5
TM2A	100	29.5	35.0	36.5	40.0
TM3	90	31.0	34.5	37.0	40.0

The results indicated that little or no additional strength had been achieved with the highest cement content. It was decided to further develop mix TM2A with varying aggregate contents to assess workability and produce an acceptable surface finish.

7.2 Final mix trials

Three panels (1.3m x 0.4m x 0.35m) were initially cast, with sand contents ranging from 40% to 46%. The workability of the mix and the surface finish of the panel assessed. The rich mix (40% sand) gave the best overall results and it was agreed that this would be further developed, varying the water content, to achieve the optimum mix for approval.

A further three panels were cast with the mix designs detailed in Table 3. 150mm cubes were also cast to ensure the mix would achieve the strength requirements (see Table 4).

Table 3. Final trial panel mix designs

Ref	Cement (kg)	Aggregate Coarse (kg)	Fine (kg)	Water (kg)	Free w/c	Admixture (l)
TM2B	350	1185	780	120.4	0.344	2
TM2C	350	1185	780	119.7	0.342	2
TM2D	350	1185	780	119.0	0.340	2

Table 4. Final trial panel compressive strength results

Ref	Slump (mm)	3 day (N/mm^2)	7 day (N/mm^2)	28 day (N/mm^2)
TM2B	85	29.5	33.0	39.5
TM2C	90	30.0	34.0	40.5
TM2D	55	30.0	35.5	40.5

The properties of these mixes were accepted as being sufficient for full trial panels to be cast.

7.3 Full trial panels

Two full trial panels, one containing reinforcing steel of a similar size and configuration to that specified for the counterweights and one unreinforced were cast using the expected production procedures, equipment and steel moulds. The quantities of constituents in the mix design were 'rounded', to give a final mix design as reported in Table 5. During the trial the cement content of the first batch was increased by 15kg/m^3 to 'butter' the mixer, shutes and barrows, etc. The strengths obtained from the various batches are summarised in Table 6.

On completion of the casting it was discovered that the plasticiser had been underdosed in batch no. 5, resulting in the low slump recorded, increased difficulty in compaction and a lower quality surface finish than expected.

A second trial panel was therefore cast, without reinforcement (no cage was available), using an identical mix design, but with careful attention to the quantity of plasticiser in order to produce a reference surface finish standard. The surface achieved by this second panel were accepted by the Client and the Engineer.

7.4 Temperature control measures and materials processing

The temperature control measures proposed by the Contractor to ensure the initial mix temperature complied with the specification proved successful during the trials. The storage of cement in air conditioned containers at a temperature of 24° to 26°C, the use of crushed ice in the mixing water, shading the fine aggregate and production facilities and cooling the coarse aggregate with water combined to keep the initial mix temperature in the order of 25°C.

Table 5. Full trial panel mix design

Ref	Cement (kg)	Aggregate Coarse (kg)	Fine (kg)	Water (kg)	Free w/c	Admixture (l)
FTP1	350	1200	815	120.0	0.343	1.8

Table 6. Full trial panel compressive strength results

Batch No.	Slump (mm)	3 day (N/mm²)	7 day (N/mm²)	28 day (N/mm²)
3	50	26.9	32.4	42.3
4	50	28.7	33.8	41.9
5	35	29.4	35.4	42.5

The as delivered coarse aggregate contained a high proportion of 'fines' (material passing the 5mm sieve. It was recommended that the proportion of this fraction be reduced to less than 4.0% of the total in order to improve the quality of the surface finish of the concrete. This was achieved by screening the aggregate, by hand, through a 5mm sieve.

8 Initial production of counter weight units

A purpose designed casting shed had been constructed for the production of the units, containing two drum mixers of 0.48m³ capacity, two concrete skips (transported by gantry crane) and the steel forms. Aggregate pens and hard standings for grit blasting and storage of reinforcement cages were also constructed.

The first units produced consisted of a beam and a capping unit, cast on consecutive days.

8.1 Temperature control of the mix constituents

In order to meet the specification requirement that no concrete be placed at ambient temperatures exceeding 30°C, production was scheduled to take place during the coolest part of the day i.e. between 4am and 8am.

The temperatures of the raw materials was as follows:

Cement   24° to 28°C
Coarse aggregate   24° to 26°C

Fine aggregate	35° to 36°C
Water	8° to 10°C

8.2 Concrete production

To prevent overloading and grout loss from the mixer batches of 0.46m³ were produced during the casting of the beam, which was reduced to 0.43m³ for the capping unit. All the solid ingredients were weigh batched. Water was batched from calibrated tanks on the mixers and was added by volume, the quantity added had to be varied to accommodate the moisture content of the coarse aggregate. The admixture was weigh batched into containers (one per batch of concrete) prior to production and added to the mix water.

Only one mixer was used for production of these first two units due to a fault in the water batching arrangements of one of the mixers.

The concrete mixer was initially "buttered", using a sand and cement mix which was discharged into wheelbarrows and discarded. The mixing procedure used for the concrete production was as follows:

(1)   Coarse aggregate, fine aggregate and cement were weigh batched into the mixer loading hopper.
(2)   The water was batched into the mixer water tank.
(3)   The plasticiser was added into the discharge valve chamber of the mix water tank.

((1), (2), and (3) took place simultaneously).

(4)   The water and plasticiser were discharged into the mixer and mixed for one to two minutes.
(5)   The aggregates and cement was discharged into the mixer and mixed for four to five minutes.
(6)   The contents of the mixer were partially discharged into the concrete skip and samples taken for cube tests and slump tests (as required) and the temperature measured.
(7)   The remaining concrete was discharged into the skips for transportation to the formwork.

The mix temperatures of the fresh concrete varied between 28° and 30.5°C (the high temperature being the final batch produced to complete the casting of the capping unit).

The concrete was placed directly into the moulds from the skips in horizontal layers less than 0.5m thick along the whole length of the formwork. Compaction was achieved by means of 54mm diameter poker vibrators. The units were finished by hand to produce a 'steel float finish'.

Curing was as described in the Contractors working procedure (Section 6).

8.3 Surface appearance

Formwork was stripped at 24 hours and the units inspected. The beam showed a number of defects including voids at the surface, sand rich areas, blow holes, colour variation and a cold joint. The appearance of the concrete was not of an equivalent standard to that of the control panel and was consequently rejected. The cause of the defects observed was attributed to three factors, inadequate screening of the coarse aggregate, variation in the water content of the concrete and the length of the unit resulting in large time periods between vertical layers being placed.

The capping unit showed a high quality, smooth surface finish with only minor imperfections. This unit compared favourably with the control panel and was accepted.

9 Full production

After the initial production problems were cured, full production has continued on a regular schedule and the works are proceeding within timetable.

10 Conclusion

High quality concrete can be made in hostile environments (very hot) with only limited resources. However, this requires careful specification of materials and procedures. Ensuring that the highest quality raw materials available are used and that the quality is maintained throughout the contract. That basic precautions are taken to restrict the temperature of the fresh concrete to reasonable levels by cooling the raw materials and that the placed concrete is prevented from heating by rigorous curing procedures.

21 DURABILITY ASPECTS OF THE KING FAHD CAUSEWAY

J. M. J. M. BIJEN
INTRON, Sittard, and Technical University of Delft,
The Netherlands

Abstract

Shortly after the contract for the causeway between Saudi Arabia and Bahrain was awarded, a durability study group was formed by the contractor and charged with the task of studying various durability aspects of the 12 km long concrete bridges. On the basis of the findings of this study group the technical specifications were substantially modified. Among the changes were the use of portland blast furnace slag cement with a high slag content (>70%) and an increase in cover.

A first test pier was placed in the Gulf in 1982. As contrasted with the piers of the bridge the concrete of this test pier was not protected by any coating. Results obtained on cores extracted from these piers and the structure itself a number of years after the first exposure are presented and interpreted, as well as some results on cores from the bridge itself.

It is concluded that the reinforcement of the bridge appears to be well preserved. Chloride penetration proceeds only very slowly.

Salt weathering of the concrete surface was observed in the splash-zone. An investigation has been carried out into the vulnerability of the portland blast furnace cement concrete applied in comparison with a reference concrete based on a sulphate resistant portland cement. The results show that in case of prolonged wet curing there is no significant difference, but the former concrete is more vulnerable to dry curing.

Keywords: Durability

1 Introduction

In May 1981 the contract for the construction of the Saudi Arabia Bahrain Causeway (later on called the King Fahd Causeway) was awarded jointly to Ballast Nedam, the Netherlands and Bandar Ballast Contractors of Saudi-Arabia. The bridge was an alternative design of Ballast Nedam Engineering. Directly after the contract was awarded the contractor formed a durability study group consisting of specialists from the contracting firm, from Messrs Sandberg of London, from Intron, institute for material and environmental research, the Netherlands and Prof.ir. P.C. Kreijger from the Netherlands. The task of the study group

Concrete in Hot Climates. Edited by M. J. Walker. © RILEM
Published by E & F N Spon, 2 - 6 Boundary Row, London SE1 8HN. ISBN 0 419 18090 7.

was to investigate all durability aspects of the 12 km long bridges of the 25 km long causeway and to provide recommendations for improvements for those aspects where to the opinion of the durability study group the specifications were insufficient to warrant a long servicelife of the bridges. The specifications for the concrete works were made by the consultant of the Ministry of Finance of Saudi Arabia, which was the client. The durability study group reported in 1982. In that period it was recognized by the experts of the study group that the concrete structures were situated in a very aggressive environment which was typical of the Gulf (see the publications of Fookes and Collis (1976) and Fookes, Pollock and Kay (1981)). The findings of the durability study group led to recommendations for major changes in the initial specifications. Most of the recommendations were accepted by the client and an order to implement these improvements was granted to the contractor. The findings and the recommendations of the durability group have been reported in a previous paper, Van Heummen et al. (1985).

The performance of the bridges was monitored among others by a special team of the World Bank and was reported to be good, Gerwick (1990). In the 10 years after the first (test) pier was placed in the Gulf, a number of tests were done. Part of these tests were carried out at the author's institute, mainly concerning surface scaling and chloride penetration. Although these tests do not involve a systematic and a statistically based investigation they do provide sufficient information as to how the bridges were performing with respect to the main degradation factors.

2 Specifications, Concrete Compositions and Curing

2.1 Specifications
In table 1 the main items regarding specifications in relation to durability are summarized. Important changes with respect to the initial specifications are:
- the use of a high slag content portland blast furnace slag cement, type HOZ 35L-NW/HS/NA meeting the German standard DIN 1164, instead of an ASTM Type V sulphate resisting portland cement,
- an increase in the cover on the reinforcement of the piers and of the bridgedeck from 45 to 75 mm and
- a maximum allowable chloride content of the concrete of 0.1% m/m by mass of cement instead of 0.3.

The durability study group was of the opinion that the protection of the reinforcement and prestress tendons must be provided for solely by the concrete and could not be relied upon the prescribed coating on the concrete of the substructure.

2.2 Concrete composition and curing
A typical composition used is shown in Table 2. The cement comprised an average of about 71% ground granulated blast furnace slag. The chloride content was less than 0.02% m/m. The coarse aggregate used was a crushed gabbro rock from various sources in the United Arab Emirates. The fine aggregate was a locally dredged marine sand consisting mainly of α-quartz and calcite. A high range water reducer (a sulphonate naphthalene

232

condensate) appeared to be essential for achieving the low water-cement ratio of 0.38. With the exception of the shafts of the main span all the other concrete elements were prefabricated.

Table 1. Main items regarding specifications with respect to durability, King Fahd Causeway

concrete	
cement type	DIN 1164 HOZ 35L-NW/HS/NA sulphate resisting portland blast furnace cement, high slag content 70-80% m/m
water cement ratio	≤ 0.42
chloride content of concrete	≤ 0.1 % m/m by mass of cement (Cl⁻)
cover	
substructure	≥ 75 mm
superstructure - deck	≥ 75 mm
- rest	≥ 50 mm
epoxy coating on concrete surface of substructure	from 2 m + MSL to 4 m + MSL*

* later extended to 8 m + MSL

Table 2. Concrete mix composition in kgs per m³

400 kgs	Portland blast furnace cement (HOZ 35L-NW/HS/NA) with a slag content of 70-80% and a maximum chloride content of 0.02% (as Cl⁻)
610 kgs	washed marine sand
565 kgs	coarse aggregates 4 -12.5 mm
807 kgs	coarse aggregates 12.5-25 mm
152 l	effective water content*
9 l	absorption water
5.5 l	superplasticizer (40% solution in water)
0.5 l	retarder

* effective water/cement ratio ≈ 0.38

Part of the concrete was steam cured and the curing regime chosen was:
- initial temperature of fresh concrete 32°C
- dormant period minimum 2 hours
- heating rate of 10°C per hour as a maximum
- maximum concrete temperature of 50-75°C, preferably 60°C
- cooling rate of 5-10°C per hour

After steam-curing the compressive strength had to be at least 75% of the minimum strength required after 28 days. Subsequent to steam-curing the concrete should be cured wet for 3 to 5 days. The concrete of the piers was only steam-cured in the winter period, while the superstructure was mostly steam-cured.

3 Surface weathering of concrete surface

During the first years after construction removal of the cement skin of the concrete (surface weathering) was observed on the piers above the coated area (up to 4 m + MSL). The cement skin appeared to be removed up to a few mm's. The surface weathering incurred here was greater than that for similar structures in Western Europe. The removal of the concrete cement skin was highest directly above the coating on the piles and strongly decreased in severity with increasing height.

The deterioration was due to salt crystallization due to the splashing of seawater on the piers and its subsequent evaporation from the piers. The splashing zone appeared to commence at about 4 m + MSL.

It was decided to extend the coating to 8 m + MSL and to start a test programme to determine the extent of the scaling and the sensitivity of the portland blastfurnace cement used in relationship to the sulphate-resisting portland cement that was initially specified.

3.1 Samples extracted from the bridges
Forty cores were drilled from various parts of the piers and the superstructure. In Table 3 the average values for the depth of carbonation are shown. Pore size distribution was measured by means of mercury porosimetry on both the surface layer and the bulk of the cores. Typical pore size distribution data are presented in Figure 1. The percentage of the pores larger than 30 nm (used as a border between gel and capillary pores) for some of the samples is given in Table 4. Chloride profiles were measured by sectioning the cores. Typical chloride concentration profiles are shown in Figure 2 for cores taken at 12, 10 and 8 m above MSL respectively. It is clear from the data that the chloride content in the surface layer tends to increase with decreasing distance from the water level.

Table 3. Average rates of depth of carbonation and cover for various parts of the superstructure, the bridge decks, the kerbs, the piers and the pier connection (in mm).

part	depth of carbonation			cover		
part	number of cores	average	standard deviation	number of cores	average	standard deviation
side of super-structure	12	6.9	1.7	9	55.9	4.1
kerbs of super-structure	6	6.0	1.3	2	58.5	-
bridge decks	10	5.4	0.5	2	78	-
piers	3	9.7	3.2	-	-	-
pier connections	9	10.4	1.9	1	87	-

Table 4. Volume percentage of pores larger than 30 nm.

concrete samples	percentage of pores > 30 nm
bulk* no. 5 (superstructure)	29
surface no. 5 (superstructure)	51
surface no. 8 (superstructure)	46
surface no. 14 (kerbs)	35
surface no. 23 (bridge deck)	46
bulk no. 28 (bridge deck)	35
surface no. 28 (bridge deck)	50
surface no. 30 (piers)	52
surface no. 31 (piers)	51
surface no. 39 (pier connections)	42
surface no. 40 (pier connections)	47

* distance more than 50 mm from the surface

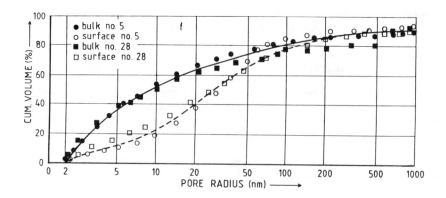

Fig. 1. Cumulative pore size distribution determined by mercury porosimetry for samples taken from the bulk and from the surface of concrete.

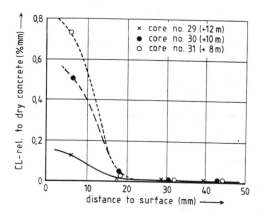

Fig. 2. Chloride profiles of cores no. 29, 30, 31 at various distances above MSL.

3.2 Accelerated exposure tests

3.2.1 Experiments

A laboratory investigation was performed with the aim of:

- assessing the effect of curing conditions on the rate of surface weathering
- determining the difference in vulnerability to surface weathering between concrete made with portland blast furnace cement and concrete made with sulphate resisting portland cement.

236

The concrete compositions used are given in Table 5. Concrete was manufactured at a temperature of 30°C. The moulding also took place at the same temperature of 30°C. The cubes with a length of 150 mm were cured in three ways. These were:

a. cured in mould at 30°C for two days while covered, after demoulding storage for 26 days at 30°C and 95% R.H.;
b. cured in mould at 30°C for two days while covered, after demoulding storage for 26 days at 30°C and 40% R.H.;
c. by steam-curing immediately after casting in a similar way as the curing on site of the King Fahd Causeway, after which the cubes were demoulded and stored at 30°C and 40% R.H. until an age of 28 days.

Table 5. Composition and some properties of the fresh concretes investigated.

concrete mix	portland blast furnace cement PBFC	sulphate-resisting portland cement SRPC
composition (kg/m³)		
cement	400	400
water, effective	152	152
water, absorbed	19	19
washed marine sand	687	695
coarse aggregate 4-12 mm	848	859
12-25 mm	444	450
Conplast 363 (high range water reducer)	1.8	1.8
	2552	2577
properties of the fresh concrete		
w/c	0.38	0.38
slump (mm)	220	195
air content (% v/v)	0.2	0.5
apparent density (kg/m³)	2557	2564
temperature (°C)	30.5	30.2

After 28 days a number of characteristics were measured. The remaining samples were subjected to the following exposure programme intended to simulate natural weathering but accelerated:
- 23 cycles of 6 hours of immersion in artificial Gulf water of 20°C followed by 66 hours at 50°C and 40% R.H. Subsequently the exposure regime was changed because no significant surface scaling was observed.
- 265 cycles of 1 hour of immersion in artificial Gulf water of 20°C followed by 6 hours at 50°C and 20% R.H. and then followed by 1 hour Gulf water immersion and lastly 16 hours storage at 50°C and 20% R.H. (hence 2 cycles a day).

3.2.2 Results
Some results of compressive strength, depth of carbonation and water penetration before

the start of the accelerated exposure and after the exposure are given in Table 6.

Table 6. 28 days compressive strength, depth of carbonation after 28 days and after 288 wet/dry cycles and water penetration according to DIN 1048.

concrete mix	curing conditions	compressive strength (MPa)	depth of carbonation (mm)		depth of water penetration	
			at the start of accelerated exposure	after accelerated exposure	average	variation
PBFC*	30°C and >95% R.H.	47	< 0,5	5-9	4	3-4
	30° and 40% R.H.	49	3-4	6-8	16	10-24
	steam-curing + 30°C/40% R.H	50	3-4	6-10	57	38-67
SRPC*	30°C and >95% R.H.	49	< 0,5	1-3	50	43-58
	30°C and 40% R.H.	47	< 0,5	3-4	66	65-67
	steam-curing + 30°C/40% R.H.	46	< 0,5	2-4	80	66-87

* PBFC = portland blast furnace cement
SRPC = sulphate-resisting portland cement

Table 7. Volume percentage of pores larger than 30 nm after 28 days of hardening.

concrete mix	curing conditions	percentage of pores > 30 nm	
		bulk	surface
PBFC	30°C and >95% R.H.	14	18
	30°C and 40% R.H.	16	37
	steam-curing + 30°C/40% R.H.	29	39
SRPC	30°C and >95% R.H.	34	43
	30°C and 40% R.H.	36	56
	steam-curing + 30°C/40% R.H.	44	55

From samples extracted from the outer 2 to 3 mm from the surface the pore size distribution was measured with mercury porosimetry. In Figure 3 a typical distribution is given for portland blast furnace cement concrete samples and in figure 4 for sulphate resisting portland cement concrete. In Table 7 the percentage of pores larger than 30 mm is given. Results of the observed rate of surface weathering after the accelerated exposure are shown in Table 8. Chloride profiles after the accelerated exposure are given in Figure 4.

Table 8. Visual inspection after 288 wet/dry cycles (9 months) of exposure.

concrete mix	curing conditions	observed extent of scaling
		wet/dry cycles
PBFC	30°C and 95% R.H.	approx. 10% concrete cement skin
	30°C and 40% R.H.	95% concrete cement skin; approx. 15% mortar skin
	s-c + 30°C/40% R.H.	100% concrete cement skin; approx. 10% mortar concrete skin
SRPC	30°C and >95% R.H.	approx. 35% concrete cement skin; approx. 15% mortar concrete skin
	30°C and 40% R.H.	approx. 5% concrete cement skin
	s-c + 30°C/40% R.H.	approx. 10% concrete cement skin

* concrete cement skin here means cement paste removed concrete mortar skin means cement + fine aggregate removed

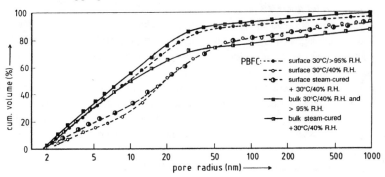

Fig. 3. Pore size distribution of samples with portland blast furnace cement.

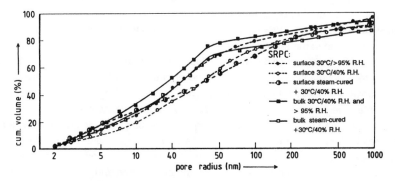

Fig. 4. Pore size distribution of samples with sulphate-resisting portland cement.

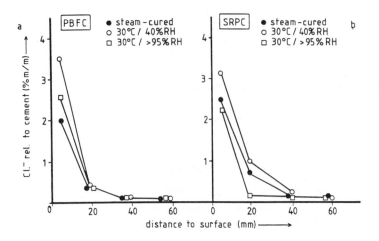

Fig. 5.   Chloride profiles after 280 wet/dry cycles: a = PBFC; b = SRPC.

3.2.3 Discussion

Both concrete mixes had the same water cement ratio of 0.38 and had the same 28 days compressive strength (of about 48 MPa).

With the same curing conditions, concrete mixes containing portland blast furnace cement tend to exhibit significantly less pores with radii larger than 30 nm compared to the corresponding concrete mixes with sulphate-resisting portland cement. This is the case both at the surface and in the bulk of the concrete.

This is likely to explain the significantly smaller maximum depth of water penetration and depth of chloride penetration observed for the former concrete mixes compared to concrete mixes with sulphate-resisting portland cement. However for the well-cured concrete, the difference in chloride penetration depth between the two concrete compositions is hardly significant.

Steam-curing followed by strong drying results in a larger percentage of pores larger than 30 nm for both types of cement.

Exposure to drying conditions at an early age results in an increase in the percentage of pores larger than 30 nm but only at the surface.

The negative effect of curing under strong drying conditions with or without previous steam curing is also reflected in the chloride-ion penetration.

For the concrete mixes containing portland blast furnace slag cement an increase in the degree of surface weathering with increasing severity of exposure to drying conditions at early age is observed. This might partly be explained by the relatively large increase (more than 100%) in the percentage of pores larger than 30 nm at the concrete surface of these mixes due to the drying conditions at an early age (18% → 39%; see table 7). The increase in larger pores is likely to be due to a considerable decrease in cement hydratation due to the drying process and to carbonation. The latter effect is known to coarsen the pore system in portland blast furnace cement, while in the case of portland cement the effects are insignificant or the opposite.

The concrete mixes containing sulphate-resisting portland cement are less sensitive to drying conditions at early age with respect to both the degree of weathering and depth of carbonation. This might be explained by the relatively small increase (less than 30%) in the percentage of pores > 30 nm at the concrete surface of these mixes due to exposure to drying conditions at an early age (43% → 56%; see table 7). However, when cured at high humidity (R.H. > 95%), concrete with portland blast furnace cement shows less weathering than the corresponding concrete mix with sulphate-resisting portland cement. This phenomenon is likely to be related to carbonation. The depth of carbonation in these experiments is greater for the portland blast furnace cement than for the sulphate-resisting portland cement concrete.

4 Chloride penetration

4.1 Samples and experiments

In 1984 and 1988 three series of cores were drilled from the piers of the causeway bridges and sent to Intron's laboratories for investigation. All series consisted of cores taken from the test pier placed as a trial in December 1982 at bridge number 4. Apart from the piers of the bridges themselves this pier in question was not coated and the concrete samples taken could be regarded as having been exposed directly to the most aggressive zone. The concrete composition of this pier was the same as that used for the bridges. The only difference is the chloride content, which for the portland blast furnace cement was higher. Later on the manufacturer changed from cooling the slag with brackish water to fresh water. Therefore the 'background' level of chloride in the concrete was about a factor two higher than that in the concrete of the bridges. A sketch of the test pier and the position of cores extracted is given in Figure 6. Core numbers 1 to 4 were drilled in December 1988.

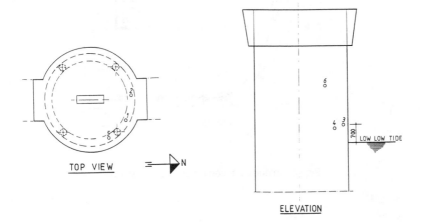

Fig. 6. A sketch of the test pier and location of cores 6-6½ years after placing.

241

The other cores were drilled in June and August 1989. The age of the concrete samples was therefore 6 to 6½ years. Of cores 5 and 6 only the distance to sea level was known. In addition to the cores from the test piers cores from the bridge piers were also analyzed. These concern the coated areas.

The cores were sectioned by dry sawing to measure the chloride profile. The carbonation depth was determined by means of the phenolphthalein test and the pore size distribution was measured by mercury porosimetry. The concrete cores were also examined with a scanning electron microscope coupled to an EDAX facility.

4.2 Results

Chloride profiles of the 6 cores extracted from the test pile are presented in Figure 7. For the coated concrete cores from the bridge piers results are given in Figure 8. The percentage of pores larger than 30 nm is given in Table 9. Table 9 shows the carbonation depth on the samples with reinforcement.

No signs of pit corrosion were detected. Occasionally some traces of rust were observed. These superficial rust traces are likely to have been present at the time of manufacturing of the concrete. In all cases the concrete cover on the reinforcement was in accordance with the specifications.

SEM microscopy and EDAX analysis confirmed that the chloride penetration was not more than 20 mm. Some crystals of ettringite were detected in entrapped air voids, see figure 9. However X-ray diffraction analysis showed the ettringite content to be rather low.

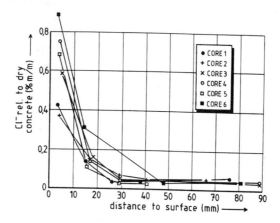

Fig. 7. Chloride-concentration profiles test pier.

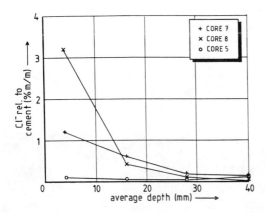

Fig. 8. Chloride profiles of cores no. 5, 7 and 8. Core no. 5 of test pier was not-coated; core no. 7 of bridge pier was coated in prefab yard; core no. 7 also of bridge pier was coated in situ after approximately one year of exposure.

Table 9. Percentage of pores larger than 30 nm as measured by mercury porosimetry and carbonation depth.

core	zone	part of core	% > 30 nm	carbonation depth	
				average	mm range
1	splash	-	-		
2	splash	bulk	19	8	5-9
2	splash	surface	51	8	6-13
3	tidal	bulk	26		
3	tidal	surface	36	1	0,5-2
4	tidal	-	-	1	0,5-2

4.3 Discussion

The cores analyzed showed chloride penetration up to a depth of about 20 nm. The profile was very steep. Samples taken in the splash zone at 5 m + MSL of the test pile showed a carbonation depth of about 8 mm, whereas in the tidal zone 2 mm was observed. From the work of Bakker it is known that in the splash zone there is a surface layer of about 15 mm which shows strong moisture fluctuations, Bakker (1991). Beyond the 15 mm line a more static moisture gradient is found. Figure 10 shows schematically this phenomenon. In the 15 mm layer, splash water is absorbed quickly and salts accumulate due to evaporation. Further penetration into the concrete beyond the 15 mm depth line proceeds much slower because the transport mechanism is diffusion of ions instead of water absorption. The accumulation of salts in the surface layer is likely to be increased by carbonation and dry

243

curing conditions. Although this phenomenon is measured in moderate climates it is believed to exist also in other marine environments, albeit with a different depth for the moisture fluctuating layer. The presence of the latter layer coincides with the fact that this outer concrete layer appears to be more permeable than the bulk of the concrete.

Taken into account also the high cover of 75 mm it can be concluded that the reinforcement is well protected and that the service life of the bridges is not endangered.

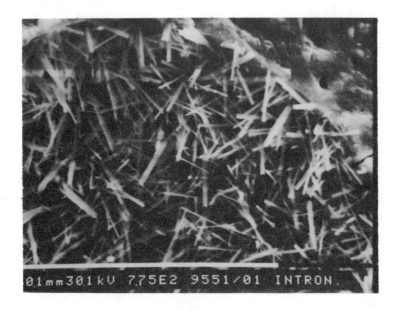

Fig. 9. SEM-photo of ettringite present in an air void.

5 Conclusions

From results of the studies carried out on more than six-year-old core samples which were fully exposed to seawater it appears that the reinforcement of the concrete of the King Fahd Causeway is well protected by the concrete cover. Accumulation of salts at the surface may occur, but from the salty surface layer migration towards the reinforcement proceeds only very slowly.

In the tests carried out the surface of wet-cured portland blast furnace cement concrete appears to weather not faster than in the case of sulphate resisting portland cement concrete. But if cured dry the portland blast furnace cement appears to be more vulnerable.

6 Acknowledgements

A debt of gratitude to Ballast Nedam of Amstelveen, The Netherlands, is acknowledged for granting permission to publish the data presented in this report.

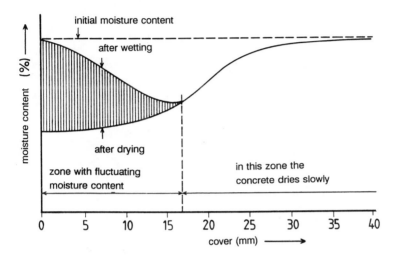

Fig. 10. Area of changing moisture content under splash water conditions according to Bakker (1991).

7 References

Bakker, R.F.M., Roessink, G. (1992), **The critical chloride content in reinforced concrete**, CUR-report.

Collis, L., Fookes, P.G., (1976), **Problems in the Middle East**, Concrete.

Fookes, P.G., Kay, E.A., Pollock, D.J., (1981), **Middle East Concrete. Rates of Deterioration**, Concrete.

Gerwick, B.C. jr., (1990) **International Experience in the Performance of Marine Cement Concrete**, Concrete International, pp. 47-53.

Van Heummen, H.,Bovee, J., Van der Zanden, J., Bijen, J., (1985) **Materials and durability**, Proceedings Symp. Saudi Arabia-Bahrain Causeway, University of Technology, Faculty of Civil Engineering, Delft, The Netherlands.

22 THE PERFORMANCE OF CATHODIC PROTECTION SYSTEMS ON REINFORCED CONCRETE STRUCTURES IN HOT CLIMATE REGIONS

W. K. GREEN
Taywood Engineering Ltd, Perth, Western Australia
D. C. POCOCK
Taywood Engineering Ltd, Southall, Middlesex, UK
S. P. LEE
Taywood Maunsell Ltd, Kowloon, Hong Kong

Abstract
Little is reported in the literature on the performance of cathodic protection (CP) systems on reinforced concrete structures in hot climate regions. This paper addresses the performance of two commercially installed systems on reinforced concrete structures in Western Australia and Hong Kong. Electrochemical performance data is discussed. A criterion using a positive shift in the base potential is suggested as an appropriate means of ensuring effective corrosion protection to reinforcement by cathodic protection.
Keywords: Cathodic Protection, Conductive Coating Anode, Mesh Anode, Discrete Rod Anode, Electrochemical Performance, Positive Base Potential Shift.

1 Introduction

The benefit of cathodic protection (CP) is proven. The passage of current from an anode through the concrete onto the reinforcement surface results in changes in the concrete environment which are beneficial to the corrosion performance of the structure including:

- The net positive current flow (electrons) towards the steel stifles the flow of positive iron ions away from the reinforcement surface.
- The potential of the reinforcement surface is polarised to values more negative than the pitting potential which restricts the dissolution of iron ions.
- At the cathode surface (the reinforcement) oxygen is reduced. This removes a depolarising component in the corrosion process.
- The oxygen reduction reaction produces hydroxyl ions at the cathode surface resulting in a surface environment which is passive to corrosion in the event that the cathodic protection system is switched off.

Concrete in Hot Climates. Edited by M. J. Walker. © RILEM
Published by E & F N Spon, 2 - 6 Boundary Row, London SE1 8HN. ISBN 0 419 18090 7.

• The passage of current will result in the electromigration of chloride ions away from the reinforcement surface. Chloride ions are known to initiate reinforcement corrosion; their removal will restore the corrosion prevention characteristics of the structure.

Cathodic protection of reinforced concrete structures has most commonly been undertaken in the USA and Europe. Little is reported in the literature on the performance of CP systems on structures in hot climate regions. This paper addresses the performance of two CP systems on reinforced concrete structures in Western Australia and Hong Kong. Electrochemical performance data is presented for 3 years of operation for the conductive coating anode CP system on the Western Australian structure. The CP system on the Hong Kong structure is a composite system comprising conductive coating anode, mesh/cementitious overlay anode and discrete rod anodes. Electrochemical data for this system at commissioning and after $2\frac{1}{2}$ months operation is detailed.

2 Commercially installed conductive coating anode systems in Western Australia

2.1 Background
Conductive coating CP systems utilising a graphite filled chlorinated rubber based paint anode have been commercially installed on two structures in Western Australia since 1988. Both structures were suffering from environmentally derived chloride ion induced reinforcement corrosion. Performance related data from one of the cathodically protected structures is presented.

The reinforced concrete walls, integral columns and free standing columns of two buildings within the structure have been cathodically protected. Installation of the system was completed in late 1988. Commissioning was conducted in January 1989.

Platinised titanium discs overlaid with strips of carbon fibre have been used as primary anodes. Protection current is supplied to four anode zones from individual 15 volt (maximum drive voltage), 1 ampere (maximum applied current) transformer-rectifier modules. Control of reinforcement potentials is achieved with embedded silver/silver chloride (Ag/AgCl) reference electrodes.

2.2 Electrochemical performance
Typical electrochemical data obtained on the conductive coating anode CP system over a period of 3 years is given in Figs 1 and 2. Two sets of electrochemical data are presented because the performance of the system varies over the structure. The results given in Fig 1 are typical of those measured at 18 of the 22 embedded reference electrode

248

locations. At the remaining 4 reference electrode locations the system performance is typical of that shown in Fig 2.

Figs 1 and 2 show the drive voltage and applied current of the transformer rectifier as well as the potential shift and 4 hour potential decay of the reinforcement. Potential shift is the potential difference between the reinforcement potential measured prior to commissioning of the CP system (i.e. base potential) and the instant-off potential. The instant-off potential is the steel potential free of any voltage gradients resulting from the protection current and resistance of the concrete. It is usually measured between 0.1 and 1.0 seconds following interruption of DC power to the anode system (Concrete Society, 1989). Potential decay is the difference between the instant-off potential and the potential after the system has been turned off for at least 4 hours.

The drive voltage results presented in Figs 1 and 2 indicate that a general rise in drive voltage was required to maintain the constant current. This suggests that the circuit resistance has increased. The resistance increase is likely to predominantly occur in two ways; the first is related to the reduction in the concentration of conducting ions (eg. chloride) in the concrete pore solution. The second is a reduction in the available voltage resulting from an increase in the Back EMF which occurs as the steel polarises. Similar trends in increasing circuit resistance were noted by Glass et al (1991) on three commercially installed systems in the UK.

The instant-off and 4 hour off potential results at Fig 1 are seen to increase with time to values more positive than the original base potential. This suggests that the passive film on the reinforcement is developed by CP and that the region of passivity is increased. The removal of chloride ions from the vicinity of the reinforcement and the increase in alkalinity from the reduction of oxygen to hydroxyl ions caused by the CP system are likely to account for these effects. Visual examination of reinforcement condition confirmed the electrochemical indications.

Glass et al (1991) report similar results for commercially installed UK systems as well as for accelerated laboratory experiments. The same effect has recently been seen on laboratory specimens following electrochemical chloride extraction treatment (see Fig 3). Corrosion rate determinations by a potentiostatic pulse polarisation resistance technique indicated passivity.

At locations where a rise in reinforcement potential to values more positive than the base potential was not observed, the commonly accepted CP criterion of a 100mV potential decay of the steel potential over 4 hours (Concrete Society, 1989) (NACE, 1990) (Manning, 1990) was met over 3 years (see Fig 2). Confirmation of corrosion protection was obtained from a visual examination of reinforcement condition. The possible reason for the measured differences in electrochemical performance at these locations is current distribution patterns. Further research into this aspect is presently underway.

It is important to note that an adjoining building within the same structure, which has not been cathodically protected, has shown a 30% increase in concrete spalling and delamination over the last 3 years due to chloride ion induced reinforcement corrosion. A conductive coating anode CP system is currently being designed for this building.

3. Commercially installed system on Mongkok Bridge, Hong Kong

3.1 Background

Mongkok Bridge, Hong Kong consists of a single span of precast post-tensioned longitudinal beams bearing on two cantilevered abutments. The reinforced concrete deck slab was cast in-situ and is overlaid with gravel ballast supporting three railway lines. The bridge was constructed in 1979 in two separate halves for practical reasons. The two halves were subsequently joined by a 'stitch strip' containing calcium chloride accelerator.

The stitch strip exhibited concrete damage after several years service due to chloride induced reinforcement corrosion. Although the damage had not affected the structural capacity of the bridge, concerns were raised on the hazard to the public from falling, spalled concrete. The bridge was thus repaired and a conductive coating anode CP system was installed in 1986.

A recent survey of the bridge revealed water leakage through cracking in the deck superstructure had impaired the performance of the conductive coating anode CP system. In addition, degradation of the protective surface coating on the abutment and minor concrete spalling were identified. Subsequently, a refurbishment of the bridge including repair and coatings works was carried out between October 1991 and February 1992.

3.2 Installation in 1986

The CP system installed in 1986 utilised a graphite filled chlorinated rubber based paint anode. Platinised titanium discs overlaid with strips of conductive carbon fibre were used as primary anodes. Monitoring of reinforcement potentials was achieved by means of embedded Ag/AgCl reference electrodes.

The system was operating satisfactorily until water leakage through the deck superstructure occurred causing blistering of the conductive coating anode at some locations.

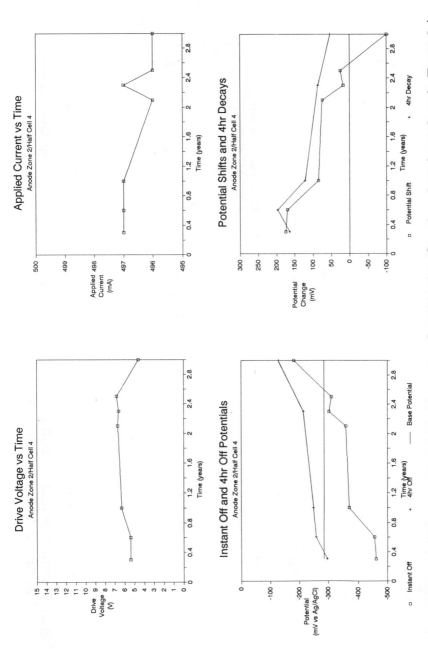

Fig.1. Electrochemical Performance of a Conductive Coating Anode at Anode Zone 2 / Location 4 on a structure in Western Australia.

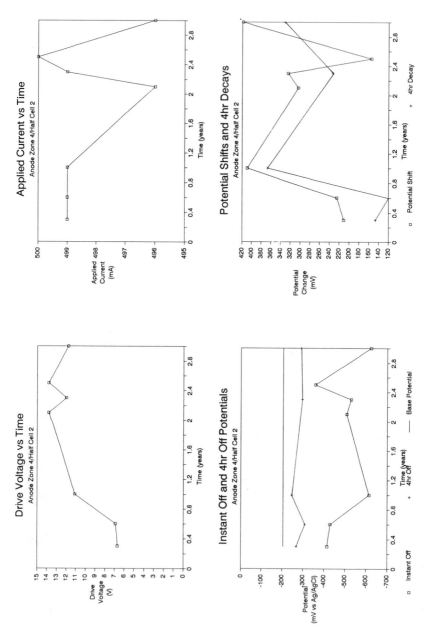

Fig.2. Electrochemical Performance of a Conductive Coating Anode at Anode Zone 4 / Location 2 on a structure in Western Australia.

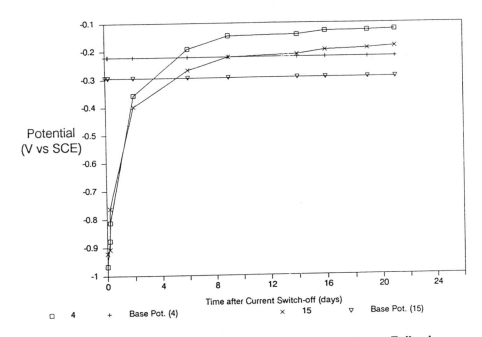

Fig.3. Potential Decay of Laboratory Specimens Following
Chloride Extraction Treatment.

3.3 Refurbished system

Results of the condition survey of the 1986 system
indicated that the conductive coating anode on the top
surface of the bridge deck was in good condition. This was
therefore retained in the refurbished CP system. The
conducting coating anode on the remaining anode zones was
replaced with a mixed metal oxide coated titanium mesh/
cementitious overlay anode system to withstand the periodic
water movement in the structure. To provide supplementary
current to deeply embedded corroding reinforcement,
additional platinum coated titanium discrete rod anodes
were installed.

Manganese/manganese dioxide (Mn/MnO_2) reference
electrodes were embedded in each anode zone to replace the
previous Ag/AgCl reference electrodes.

The refurbished system was commissioned in March 1992.

3.4 Electrochemical performance of the refurbished system

The potential shift and potential decay results at commissioning and after approximately $2\frac{1}{2}$ months of operation for the conductive coating anode on the top surface of the bridge deck are shown at Fig 4. The same results for the mesh anode and discrete rod anodes are shown at Figs 5 and 6 respectively. The drive voltage and applied current of the transformer rectifier modules for the respective anode systems at commissioning and after $2\frac{1}{2}$ months are summarised at Table 1 below.

Table 1. Drive voltage and applied current for Mongkok Bridge
CP system

	Coating Anode Anode Zones 1 to 3	Mesh Anode Anode Zones 4 to 7	Discrete Anodes Anode Zones 5 & 7 Locations A1 & A2
Commissioning Drive Voltage (V) Applied Current (mA)	5.6 700	2.5 4.2 3.0 3.2 699 704 701 694	2.5 2.4 159 152
$2\frac{1}{2}$ months Operation Drive Voltage (V) Applied Current (mA)	5.3 402	2.5 6.2 2.9 4.7 401 384 402 370	2.8 3.2 86 86

The protection criterion adopted for the Mongbok Bridge CP system was a minimum 150mV potential shift and a 100mV potential decay over approximately 4 hours depolarisation. Figs 4 and 5 show that a 150mV potential shift was achieved at nearly all anode zones for the conductive coating anode and mesh anode sections of the CP system at commissioning and after $2\frac{1}{2}$ months of operation.

254

A 100mV potential decay was however only achieved at two of the four mesh anode zones. For the conductive coating anode zones potential decay values of less than 100mV were achieved particularly at $2\frac{1}{2}$ months system operation. Future monitoring and adjustment visits will determine whether it is possible, or necessary, to achieve a 100mV potential decay at these locations.

It is interesting to note that 7 day potential decay values measured at $2\frac{1}{2}$ months system operation were more positive than base (i.e. pre-CP commissioning) potentials at 5 of the 12 conductive coating monitoring locations and 12 of the 16 mesh anode monitoring locations (see Table 2). It is intended that achievement of potential decay results more positive than base potential values will be adopted as the primary criterion for evaluating the protection performance of the CP system.

The potential shift and potential decay results at Fig 6 for the discrete rod anodes indicate that it has as yet not been possible to achieve 150mV potential shifts and 100mV potential decays. However, the 7 day potential decay values after $2\frac{1}{2}$ months system operation (see Table 2) were all considerably more positive than base potentials values. This suggests that corrosion protection is being provided to the deeply embedded reinforcement by the discrete rod anodes.

The applied current results at Table 1 show that an upto 80% reduction in cathodic current could be effected after $2\frac{1}{2}$ months of operation of the system. Potential shift and potential decay values were not significantly lowered hence the corrosion protection provided by the CP system was not compromised. The life of the CP system will therefore be maximised. Acidic substances are produced at the surface of the anode as a result of the anode reactions. The rate of acid production is proportional to the magnitude of the applied current. It is this acid production which determines the useful life of the anode.

Table 2. 7 day potential decay values for Mongkok Bridge
CP system after 2½ months operation

Anode Zone/Location	Base Potential (mV vs Mn/MnO$_2$)	7 day Potential Decay (mV vs Mn/MnO$_2$)
Conductive Coating Anode		
1A1	−308	−329
1A2	−312	−333
1B1	−323	−289
1B2	−321	−294
2A1	−182	−217
2A2	−191	−211
2B1	−176	−200
2B2	−179	−202
3A1	−292	−236
3A2	−293	−233
3B1	−273	−170
3B2	−272	−167
Mesh Anode		
4A1	−314	−372
4A2	−298	−366
4B1	−366	−268
4B2	−358	−274
5B1	−282	−289
5B2	−282	−336
5C1	−276	−220
5C2	−273	−203
6A1	−273	−111
6A2	−266	−125
6B1	−320	−159
6B2	−316	−213
7B1	−481	−227
7B2	−466	−224
7C1	−303	−268
7C2	−293	−269
Discrete Anodes		
5A1	−279	−124
5A2	−278	−123
7A1	−223	−107
7A2	−224	−101

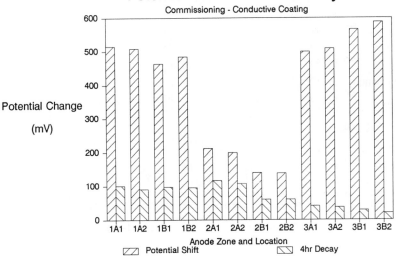

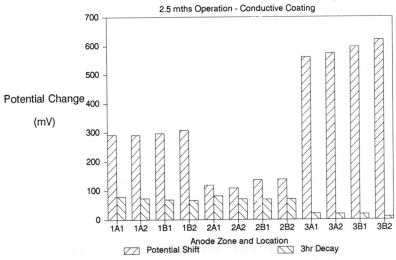

Fig.4. Potential Shift and Potential Decay results for the Conductive Anode on Mongkok Bridge, Hong Kong, at commissioning and after 2.5 months operation.

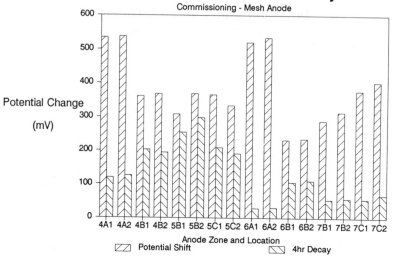

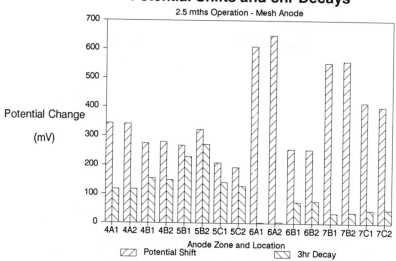

Fig.5. Potential Shift and Potential Decay results for the Mesh Anode on Mongkok Bridge, Hong Kong, at commissioning and after 2.5 months operation.

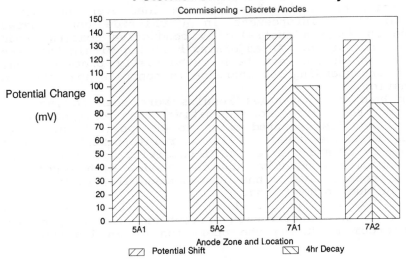

Potential Shifts and 4hr Decays

Commissioning - Discrete Anodes

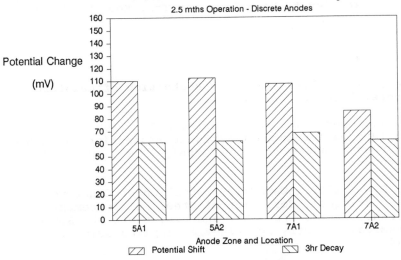

Potential Shifts and 3hr Decays

2.5 mths Operation - Discrete Anodes

Fig.6. Potential Shift and Potential Decay results for the
Discrete rod Anode on Mongkok Bridge, Hong Kong, at
commissioning and after 2.5 months operation.

5 Conclusions

1. Corrosion protection has been provided over the last 3 years to the reinforcement in a chloride contaminated building in Western Australia by conductive coating anode cathodic protection. An adjoining building within the same structure which has not been cathodically protected has shown a corresponding 30% increase in concrete spalling and delamination.
2. Four hour off potential results were seen to increase with time to values more positive than original base potential results measured prior to CP application. This suggests that the passive film on reinforcement has been developed by CP and the region of passivity has increased.
3. A composite conductive coating anode, mesh anode and discrete anode CP system has been successfully installed on Mongkok Bridge in Hong Kong.
4. A positive base potential shift could be adopted as a criterion to ensure corrosion protection is being provided to reinforcement during the operation of an installed CP system.

6 Acknowledgements

The authors wish to thank the Directors of Taywood Engineering for permission to publish this paper. Special thanks are extended to Dean Rowe and Tony Keeping for their efforts.

7 References

Concrete Society, 1989, **Cathodic Protection of Reinforced Concrete,** Technical Report No. 36.

Glass, G.K., Green W.K. and Chadwick J.R., (1991), **'Long Term Performance of Cathodic Protection Systems on Reinforced Concrete Structures',** UK Corrosion 91 Conference, Manchester.

Manning, D.G., 1990, 'Cathodic Protection of Concrete Highway Bridges', **Corrosion of Steel Reinforcement,** ed. C.L. Page, K.W.J. Treadaway and P.B. Bamforth, Society of Chemical Industry and Elsevier Applied Science, 486-497.

NACE, 1990, **'Cathodic Protection of Reinforcing Steel in Atmospherically Exposed Concrete Structures',** Standard Recommended Practice RP0290-90 Item No. 53072.

AUTHOR INDEX

SUBJECT INDEX

This index has been compiled from the keywords assigned to the individual papers, edited and extended as appropriate. The numbers refer to the first page number of the relevant paper.

RILEM, The International Union of Testing and Research Laboratories for Materials and Structures, is an international, non-governmental technical association whose vocation is to contribute to progress in the construction sciences, techniques and industries, essentially by means of the communication it fosters between research and practice. RILEM activity therefore aims at developing the knowledge of properties of materials and performance of structures, at defining the means for their assessment in laboratory and service conditions and at unifying measurement and testing methods used with this objective.

RILEM was founded in 1947, and has a membership of over 900 in some 80 countries. It forms an institutional framework for cooperation by experts to:

* optimise and harmonise test methods for measuring properties and performance of building and civil engineering materials and structures under laboratory and service environments;

* prepare technical recommendations for testing methods;

* prepare state-of-the-art reports to identify further research needs.

RILEM members include the leading building research and testing laboratories from around the world, industrial research, manufacturing and contracting interests as well as a significant number of individual members, from industry and universities. RILEM's focus is on construction materials and their use in buildings and civil engineering structures, covering all phases of the building process from manufacture to use and recycling of materials.

RILEM meets these objectives though the work of its technical committees. Symposia, workshops and seminars are organised to facilitate the exchange of information and dissemination of knowledge. RILEM's primary output are the technical recommendations. RILEM also publishes the journal *Materials and Structures* which provides a further avenue for reporting the work of its committees. Many other publications, in the form of reports, monographs, symposia and workshop proceedings, are produced.

Details of RILEM membership may be obtained from RILEM, École Normale Supérieure, Pavillon du Crous, 61, avenue du Pdt Wilson, 94235 Cachan Cedex, France.

Details of the journal and the publications available from E & F N Spon/Chapman & Hall are given below. Full details of the Reports and Proceedings can be obtained from E & F N Spon, 2-6 Boundary Row, London SE1 8HN, Tel: (0)71-865 0066, Fax: (0)71-522 9623.

Materials and Structures

RILEM's journal, *Materials and Structures*, is published by E & F N Spon on behalf of RILEM. The journal was founded in 1968, and is a leading journal of record for current research in the properties and performance of building materials and structures, standardization of test methods, and the application of research results to the structural use of materials in building and civil engineering applications.

The papers are selected by an international Editorial Committee to conform with the highest research standards. As well as submitted papers from research and industry, the Journal publishes Reports and Recommendations prepared buy RILEM Technical Committees, together with news of other RILEM activities.

Materials and Structures is published ten times a year (ISSN 0025-5432) and sample copy requests and subscription enquiries should be sent to: E & F N Spon, 2-6 Boundary Row, London SE1 8HN, Tel: (0)71-865 0066, Fax: (0)71-522 9623; or Journals Promotion Department, Chapman & Hall, 29 West 35th Street, New York, NY 10001-2291, USA, Tel: (212) 244 3336, Fax: (212) 563 2269.